CW01431300

Rivalry in the Information Sphere

Russian Conceptions of Information Confrontation

MICHELLE GRISÉ, ALYSSA DEMUS, YULIYA SHOKH, MARTA KEPE, JONATHAN W. Welburn, KHRYSTYNA HOLYNSKA

Prepared for the U.S. European Command
Approved for public release; distribution unlimited

RAND NATIONAL DEFENSE RESEARCH INSTITUTE

For more information on this publication, visit **www.rand.org/t/RRA198-8**.

About RAND

The RAND Corporation is a research organization that develops solutions to public policy challenges to help make communities throughout the world safer and more secure, healthier and more prosperous. RAND is nonprofit, nonpartisan, and committed to the public interest. To learn more about RAND, visit www.rand.org.

Research Integrity

Our mission to help improve policy and decisionmaking through research and analysis is enabled through our core values of quality and objectivity and our unwavering commitment to the highest level of integrity and ethical behavior. To help ensure our research and analysis are rigorous, objective, and nonpartisan, we subject our research publications to a robust and exacting quality-assurance process; avoid both the appearance and reality of financial and other conflicts of interest through staff training, project screening, and a policy of mandatory disclosure; and pursue transparency in our research engagements through our commitment to the open publication of our research findings and recommendations, disclosure of the source of funding of published research, and policies to ensure intellectual independence. For more information, visit www.rand.org/about/principles.

RAND's publications do not necessarily reflect the opinions of its research clients and sponsors.

Published by the RAND Corporation, Santa Monica, Calif.
© 2022 RAND Corporation
RAND® is a registered trademark.

Library of Congress Cataloging-in-Publication Data is available for this publication.

ISBN: 978-1-9774-0717-7

Cover: Mlenny/Getty Images.

About This Report

The informatization of societies in the 21st century has led to the emergence of the concept of information confrontation [*informatsionnoe protivoborstvo*, or IPb] as a component of Russian strategic thinking. Based on analysis of the Russian military-scientific literature and Russian strategic documents, this report introduces the concept of information confrontation, including prevailing definitions and existing typologies with the Russian military-scientific literature, discusses the historical evolution of the concept, analyzes the role of IPb in Russian national security and defense strategy, and assesses Ukrainian scholarly discourse on Russian IPb campaigns targeting Ukraine.

The research reported here was completed in April 2021 and underwent security review with the sponsor and the Defense Office of Prepublication and Security Review before public release.

RAND National Security Research Division

This publication was funded by the Russia Strategic Initiative, United States European Command, and conducted within the International Security and Defense Policy Center of the RAND National Security Research Division (NSRD), which operates the National Defense Research Institute (NDRI), a federally funded research and development center sponsored by the Office of the Secretary of Defense, the Joint Staff, the Unified Combatant Commands, the Navy, the Marine Corps, the defense agencies, and the defense intelligence enterprise.

For more information on the RAND International Security and Defense Policy Center, see www.rand.org/nsrd/isdp or contact the Center director (contact information provided on the webpage).

Contents

Figures and Tables

Summary

The role of information and information technologies in strategic competition and military operations has evolved considerably in the first two decades of the 21st century. This evolution has had a profound impact on the Russian military-scientific community and its perceptions of the relationship between information and conflict. In the 21st century, information and information technologies infuse all aspects of society—in peacetime, during periods of competition, and during wartime. Since the early 2000s, Russian military experts have observed that information technologies complicate and effect the course of military conflicts, which often have their origins in events that occur in the information space.

The Russian military-scientific literature has frequently noted that the rise of advanced information technologies, which allow for the rapid sharing, processing, and analysis of data, has had a significant effect on the character of military operations. As one Russian military scholar notes, "Information wars are never cold, but in these wars, people are programmed, not killed."[1] Information has played an important role in shaping the course of strategic competition and military conflict, and in influencing thinking within the Russian military-scientific community on related issues. The transformative effect of information has led to the emergence of the concept of information confrontation [*informatsionnoe protivoborstvo*, or IPb] as a component of Russian strategic thinking. Over the last two decades,

---

[1] Sergey M. Pyastolov, "Slovo Kak Oruzhiye [Word as a Weapon]," *Informatsionniye Voyny [Information Wars Journal]*, Vol. 1, No. 49, 2019, pp. 17–21.

the Russian military-scientific community has addressed at length the defining features, historical antecedents, and present-day role of IPb as a means of strategic competition. In recent years, moreover, the execution of IPb campaigns in Ukraine has led to a parallel discussion of this concept within the Ukrainian military-scientific literature. The Russian invasion of Ukraine in February 2022 will likely lead to new insights and prompt new discussion within both the Russian and Ukrainian military-scientific communities regarding the practice of IPb during conflict.

This report introduces the concept of information confrontation, including prevailing definitions and existing typologies with the Russian military-scientific literature. It provides a discussion of information confrontation in historical context, describing the evolution of influence operations and psychological warfare from the late 18th century to the Vladimir Putin era. It examines the role of IPb in Russian national security and defense strategy, analyzing how major strategic documents address IPb and related concepts. In addition, this report analyzes the discourse within the Ukrainian military-scientific community on Russian information confrontation campaigns targeting Ukraine.

Findings

Our analysis revealed that while the concept of IPb is discussed at length in the Russian military-scientific literature, there is not a coherent, unified doctrine of information confrontation, nor is there a standardized definition of IPb or related concepts within this literature or within Russian strategic documents. For over two decades, scholars within the Russian military-scientific community have called for the standardization of key terms, concepts, and definitions related to information confrontation and information warfare. At the same time, the Russian military-scientific community has closely followed developments in U.S. military doctrine and strategy pertaining to the information domain, with Russian military scholars drawing lessons from U.S. and coalition military operations since the end of the Cold War. The Gulf War, in particular, has played a profound role in shaping Russian perceptions of and approaches to IPb—as well as modern warfare more broadly—by demonstrating the effectiveness of using informational means to influence the course of a military conflict and the possibility of engaging in what is often referred to as "contactless war."

Our analysis also reveals that the Russian military-scientific literature, as well as the Russian literature on international law, frequently equates information weapons with weapons of mass destruction (WMD), viewing both types of weapons as having the potential to effect massive changes within the international system. The evolution of IPb—from something that is primarily carried out to supplement conventional military operations during wartime to something that is carried out continuously and in peacetime—has prompted repeated calls for more effective international governance of information weapons and activities associated with information confrontation. This evolution is also apparent in our analysis of Ukrainian perceptions of Russian information confrontation.

In recent decades, information confrontation has gradually become a distinct and pervasive element of strategic competition and warfare. In 2011, Russia published the *Conceptual Views on the Activities of the Armed Forces of the Russian Federation in the Information Space*, which made apparent that Russia had begun to treat information as a military domain by delineating specific military activities, concepts, and systems associated with the information space.[2] Discourse within the Russian military-scientific literature underscores that IPb is not going to go away. Moreover, the growing role of information in all aspects of politics and society means that the more technologically advanced a country is, the more vulnerable it will be to the effects of information confrontation.[3]

---

[2] Ministry of Defense of the Russian Federation, *Kontseptual'nye Vzglyady na Deyatel'nost' Vooruzhennykh Sil Rossiyskoy Federatsii v Informatsionnom Prostranstve [Conceptual Views on the Activities of the Armed Forces of the Russian Federation in the Information Space]*, 2011, section 1.

[3] In this context, "technologically advanced" refers to countries that are more reliant on technology and whose citizens are more frequent users of information technologies.

Recommendations

We offer the following recommendations:

- First, the U.S. intelligence community can study the Russian military-scientific literature to better understand Russian activities, intentions, and perceptions in the information domain. In addition, the intelligence offices and organizations within the U.S. military services can work more closely with the Open Source Enterprise to understand publicly available and unclassified Russian-language sources and encourage the use of the Russian military-scientific journals and Russian academic journals identified in this report. The military services can also more closely monitor Russian rhetoric regarding the distortion of historical facts, especially in Eastern Europe, as an emergent tool of information confrontation. Given the continuous nature of information confrontation as an element of Russian strategy, this analysis would assist in monitoring Russian influence operations and related activities.
- Second, additional research is needed to better understand how information confrontation is operationalized in hybrid warfare, the prospects for effective international governance of the information domain, and the ways information confrontation can be used as an instrument of soft power.
- Finally, the Ukrainian military-scientific literature can similarly provide insight into Ukrainian conceptions of information confrontation. In particular, this literature offers a window into ideological shifts in Ukraine in the post-Maidan era. Studying Russia's recent implementation of this concept in Ukraine, and understanding the evolution of Ukrainian thinking on and responses to such campaigns, can ultimately help policymakers develop mitigation strategies to counter future Russian IPb activities in Ukraine and elsewhere.

Acknowledgments

We would like to acknowledge the sponsorship and support of Ken Stolworthy, Director of the Russia Strategic Initiative, U.S. European Command. Within our sponsor's office, we would like to thank Colonel Phil Forbes, United States Air Force, for his support and assistance.

We would also like to acknowledge the RAND colleagues who offered their expertise, support, and guidance throughout this effort. In particular, we would like to acknowledge and thank those responsible for leading this effort, including Mark Cozad, Dara Massicot, and Clint Reach. We also thank Michael Schwille and Seth Jones, who provided helpful feedback on an earlier draft of this report. Finally, we would like to thank Christine Wormuth, Michael Spirtas, and Francisco Walter for assisting us in making this document come to fruition.

Abbreviations

APT   advanced persistent threat

C2   command and control

C2W   command and control warfare

C3CM   command, control, and communication countermeasures

CJCS   Chairman of the Joint Chiefs of Staff

DoD   Department of Defense

EW   electronic warfare

FSB   Federal'naya Sluzhba Bezopasnosti [Federal Security Service]

GRU   Glavnoye Razvedyvatel'noye Upravleniye [Main Intelligence Directorate]

IPb   informatsionnoe protivoborstvo [information confrontation]

ISD   Information Security Doctrine of the Russian Federation

ITMO   Information Technology, Mechanics, and Optics University

MoD   Ministry of Defense of the Russian Federation

NATO   North Atlantic Treaty Organization

NSS   National Security Strategy of the Russian Federation

PsO   psikhologicheski operatsii [psychological operations]

RCB   radiological, chemical, and biological

RMD   Russian Military Doctrine

SBU   Sluzhba Bezpeky Ukrainy [Security Service of Ukraine]

WMD   weapons of mass destruction

ZVIK   Zarubezhnaya Voennaya Informatsiya i Kommunikatsiia

CHAPTER ONE

Introduction

Information has played an integral role in both strategic competition and military conflict in the first two decades of the 21st century. As some scholars have pointed out, "everything has become informational."[1] This increasing prominence of information as a core tenet of conflict and competition has caught the attention of the Russian military-scientific community, leading to the emergence of the concept of information confrontation [*informatsionnoe protivoborstvo*, or IPb] as a component of Russian strategic thinking in the post–Cold War era. Over the last two decades, the Russian military-scientific community has written at length on the defining features, historical antecedents, and present-day role of IPb as a means of strategic competition.[2] As we detail in this report, Russian military-scientific thinkers define and assess the role of IPb in varying ways, but we offer the following basic definition in an effort to lend some clarity to an otherwise complex and nebulous topic:

> Information confrontation refers to the purposeful use of offensive or defensive informational means to achieve political, economic, military, and other objectives during peacetime, competition, and wartime.

---

[1] V. I. Orlansky, "Informatsionnoye Oruzhiye i Informatsionnaya Bor'ba: Real'nost' i Domysly [Information Weapons and Information Warfare: Reality and Speculation]," *Voennaya Mysl' [Military Thought]*, No. 1, 2008, p. 62.

[2] Readers who wish to explore key articles from the Russian military-scientific literature on this topic may see Michelle Grisé, Yuliya Shokh, Khrystyna Holynska, and Alyssa Demus, *Russian and Ukrainian Perspectives on the Concept of Information Confrontation: Translations, 2002–2020*, Santa Monica, Calif.: RAND Corporation, 2021.

The role of information and information technologies in strategic competition and military operations has evolved considerably in the last 20 years. In the early 2000s, some strategists observed that modern military conflicts began in the information space and were enabled by modern information technologies.[3] They noted the growth of the internet, which had become a tool that could shape public opinion and influence political, economic, and military decisionmaking; the internet could also impact the information resources of an adversary and spread disinformation.[4] By 2008, discussions of advanced computer technologies that would have a "significant effect on the character of [military] operations in the near future" and transform conventional military conflicts into "one big 'information war'" emerged.[5] New information technologies would "increase the volume, precision, and speed of sharing, processing, and analyzing data."[6]

Through an extensive review of Russian literature, this report answers the following questions:

- What is the Russian definition of *information confrontation*?
- What are its subtypes and which Russian organizations are involved in carrying out information confrontation?
- How has information confrontation evolved over time in Russian strategy, from Imperial Russia up to Russia's 2014 campaign in Ukraine and beyond?
- How might the concept evolve in the future?

---

3 Yu. O. Yashchenko, "Internet i Informatrionnoye Protivoborstvo [Internet and Information Confrontation]," *Voennaya Mysl' [Military Thought]*, 2003, pp. 72–78.

4 Yashchenko, 2003, p. 72.

5 N. A. Molchanov, "Informatsionniy Potentsial Zarubezhnykh Stran kak Istochnik Ugroz Voyennoy Bezopasnosti RF [Information Potential of Foreign Countries as a Source of Threats to Military Security of the Russian Federation]," *Voennaya Mysl' [Military Thought]*, No. 10, 2008, pp. 2–9.

6 Molchanov, 2008, p. 8.

Methodology, Scope, and Limitations

To understand the concept of information confrontation and its role in strategic competition and modern warfare, we reviewed over 100 unclassified documents, including the Russian Military Doctrine (RMD), articles from the Russian and Ukrainian military-scientific literatures, encyclopedias and dictionaries, press releases from the Ministry of Defense of the Russian Federation (MoD), public statements by Russian leaders, and Russian media sources. These documents consisted primarily, although not exclusively, of Russian-language sources and were identified through searching databases of Russian scholarly articles, reference mining, and searches for Russian leaders and scholars who are known to have published or spoken publicly on IPb and information warfare. Our analysis included sources published or released since the end of the Cold War, although we focused on scholarship from the last two decades.

This report cites heavily from articles published in prominent Russian military-scientific journals, including *Military Thought* (Voennaya Mysl'), Russia's most prestigious military journal, and other leading military journals, including the *Journal of the Academy of Military Sciences* (Vestnik Akademii Voennykh Nauk), *Army Digest* (Armeiskii Sbornik), *Military-Industrial Courier* (Voenno-Promyshlennii Kur'er), and the *Information Wars Journal* (Informatsionniye Voyny), as well as *Flag of the Motherland* (Flag Rodiny), *Red Star* (Krasnaya Zvezda), *Battle Watch* (Boevaya Vakhta), and collections published by Russian universities. The analysis of Ukrainian perceptions of Russian information confrontation relies on articles published in the Ukrainian MoD-affiliated journal *Science and Defense* (Nauka i Oborona); *Current Issues of International Relations* (Aktual'ni Problemy Mizhnarodnyh Vidnosyn), a collection of political science works collated and published by the Institute of International Relations at the Taras Shevchenko National University of Kyiv; the *Journal of the Taras Shevchenko National University of Kyiv:Military-Special Sciences* (Visnyk Kyyivs'koho Natsional'noho Universytetu Imeni Tarasa Shevchenka: Viys'kovo-Spetsial'ni Nauky); and the *Journal of Strategic Priorities* (Zhurnal Stratehichni Priorytety), also of the Taras Shevchenko National University of Kyiv. Further, the research team examined conference papers and proceedings from national and international conferences and roundtables convened by Ukrainian institutions in the years following the events of 2014.

We deliberately chose to focus on peer-reviewed articles published by recognized experts in the field. In addition to drawing on our own understanding of prominent figures in the Russian military-scientific community, we also considered the institutional affiliations and military ranks and titles of authors, as well as the extent to which their ideas were cited by Russian officials. That said, it is important to note that these authors' prominence in the military-scientific community may not necessarily translate to actual decisionmaking authority or influence within the Russian military or other state structures. As expert Dmitry (Dima) Adamsky cautions,

> Russian strategic tradition often makes a disconnect between the words of theoreticians and the deeds of the practitioners implementing them. Military and nonmilitary theoreticians can be very advanced in their conceptualizations, but the system, as a whole can be pathologically bad at implementing them. This trait . . . has manifested throughout Russian history. . . . Thus, despite leaders' holistic approaches to strategic theory and operational planning, in reality we often observe system breakdowns (sistemnii sboi) on the operational level.[7]

This is to say that while the themes expressed in the military-scientific literature serve as an important barometer of the extent to which certain issues preoccupy the Russian military-scientific community, we cannot say with any certainty whether or how the ideas of any given author have influenced Russian decisionmaking at the highest echelons of power.

In our initial exploratory efforts to survey the landscape of writing on the subject, we found the secondary Western literature devoted to the concept of information confrontation to be relatively slim. Analyses of real-world Russian influence and information activities through an

---

[7]   Dmitry Adamsky, "Continuity in Russian Strategic Culture: A Case Study of Moscow's Syria Campaign," *Security Insights*, Marshall Center, No. 048, February 2020.

operational lens, by contrast, occupy a prominent place within the secondary literature. As such, we chose to focus our analytical efforts on the former—that is, closely examining the evolution of Russian theory on information confrontation in the hope that any insights gained from this work may prove useful in understanding Russian operations in the information environment. This report therefore focuses on the Russian theory of IPb, although we weave in operational examples where relevant in an effort to concretize this conceptual analysis.

Report Organization

The remainder of this report is organized as follows. Chapter Two discusses information confrontation in historical context, describing the evolution of information operations and psychological warfare from the late 18th century to the Putin era. Chapter Three examines the role of IPb in Russian national security and defense strategy, with a discussion of how major strategic documents address IPb and related concepts. Chapter Four uses Ukraine as a case study to examine how IPb is carried out in practice. Chapter Five presents our conclusions and recommendations. Related terms and concepts are covered in the Appendix.

Defining Information Confrontation

First, it is important to introduce the terms *information sphere, information space*, and *information environment* as they are used throughout this report. We note that the use of these terms in the Russian military-scientific literature is not always consistent with U.S. or Western understandings of these concepts. Table 1.1 defines these terms according to their usage in the context of the Russian military-scientific literature and Russian doctrine. To the extent possible, Table 1.1 also offers the definition of each term in U.S. military doctrine.

In this report, where possible, we use the term *information sphere*, as it most closely resembles the U.S. concept of *information environment*.

Table 1.1
Definitions of Foundational Terms

Term	Russian Definition	U.S. Definition
Information sphere	Aggregate of information, information infrastructure, and subjects that collect, organize, distribute, and use information, as well as systems that regulate the social relations that arise during such actions (MoD, *Military Encyclopedic Dictionary*, undated b)	No formal definition
Information space	Sphere of activity associated with the formation, creation, transformation, transmission, use, and storage of information that affects individual and public consciousness, information infrastructure, and information itself (MoD, *Conceptual Views*, 2011)[a]	No formal definition
Information environment	No formal definition, but this term is most closely related to the Russian term *information sphere*	The aggregate of individuals, organizations, and systems that collect, process, disseminate, or act on information (Joint Publication 3-13, 2014)

SOURCES: Ministry of Defense of the Russian Federation, *Military Encyclopedic Dictionary*, undated b; MoD, *Conceptual Views on the Activities of the Armed Forces of the Russian Federation in the Information Space*, 2011; Joint Chiefs of Staff, Joint Publication 3-13, 2014.

[a] Chapter Three will discuss in more detail the Russian perspective of *information space* as another military domain with its own military activities, concepts, and systems.

When citing specific sources, however, we use the terms used in those sources.

In spite of the pervasive nature and critical role of information in modern society, and the stated importance of IPb as an element of Russian national security and defense strategy, there is disagreement among experts within the Russian military-scientific community about the definition of the term. Our review of the Russian military-scientific literature and Russian strategic documents also revealed several terms and concepts that are often used in conjunction with the study and analysis of information confrontation. While some of these terms, such as *information influence*, are used to describe certain elements of information confrontation, others, such as *information war-*

fare, are frequently used synonymously with the term; still others, such as *information war*, appear as the subject of great debate in the literature on information confrontation. Nonetheless, our analysis of the literature suggests that these related terms are distinct from the concept of information confrontation and have their own differential meanings. Table 1.2 lists and defines the most frequently used related terms. The Appendix at the end of this report contains additional related terms, their definitions, and a discussion of how they are used in the military-scientific literature.

The *Military Encyclopedic Dictionary* of the Russian Ministry of Defense provides insight into the official definition of information confrontation. It defines IPb as a "form of conflict between opposing sides (states, social-political movements and organizations, armed forces, etc.), each of which seeks to defeat (inflict damage on) the enemy through informational effects in the information sphere (aggregate of information, information infrastructure, and subjects that collect, organize, distribute, and use information, as well as systems that regulate the social relations that arise during such actions), while resisting or reducing such effects on one's own side."[8] The dictionary further notes that IPb has occurred throughout history, in both wartime and in peacetime. Since the second half of the 20th century, however, the scale, content, and modes of IPb have experienced significant changes because of the "expansion of social activism and mass political participation, on the one hand, and the turbulent development of information equipment and technology, on the other [hand]."[9] The dictionary explains:

> Information confrontation is becoming an independent type of warfare in which information is appearing as a resource, means, and objective. In war, information resources and means become specialized tools for defeating the enemy, [and] depriving [the

---

[8]   Ministry of Defense of the Russian Federation, "Informatsionnoe Protivoborstvo [Information Confrontation]," *Voyennyy Entsiklopedicheskiy Slovar' [Military Encyclopedic Dictionary]*, undated c.

[9]   Ministry of Defense of the Russian Federation, undated c.

Table 1.2
Main Terms Related to Information Confrontation Used in the Literature

Category	Related Term	Definition
Type of conflict	Information war	This term is defined in the following ways: • "Confrontation between two or more states in the information space with the purpose of causing damage to information systems, processes and resources, critical and other infrastructure, undermining the political, economic and social systems, massive psychological manipulation of the population to destabilize the state and society, as well as coercing the state to make decisions in the interest of the opposing force" (MoD, *Conceptual Views*, 2011) • "Transparent and severe clash between states" characterized by causing "harmful impact on the information sphere" (MoD, *Military Encyclopedic Dictionary*, undated b) • Struggle between opposing sides for superiority over the enemy in timeliness, assurance, completeness of information, speed and quality of its processing and dissemination (Nuzhdin, 2000) • Use of "aggressive information influence" (MoD, *Defense Minister Sergey Shoygu*, 2019)
Type of warfare	Information warfare	Activities undertaken to gain information superiority in the process of armed confrontation (Rodionov, 1998)
Main activity (wartime)	Information operations	Set of information activities coordinated in terms of purpose, objects, place, and time and conducted to gain and maintain information superiority over the enemy or reduce the enemy's information superiority in a given combat theater or strategic direction (Rodionov, 1998)
Means of warfare	Information weapons	Systems and means intended for gaining (collecting) information, defeating information resources of the opposing side, and defending one's own information resources by influencing information and its carriers. The definition includes the following means of information warfare: • Technical intelligence • Mass information • Information protection • Psychotonic means • Certain types of nonlethal weapons (MoD, *Military Encyclopedic Dictionary*, undated b)
Supporting activity (peacetime)	Information influence	Peacetime activities, often conducted before the start of combat operations, undertaken to support information warfare activities (Rodionov, 1998)
Goal of conflict	Information superiority	State of friendly command and control forces and equipment having more complete, precise, validated, and timely information about the operational environment than the enemy (Grudinin, 2011)

SOURCES: Rodionov, 1998; Ministry of Defense of the Russian Federation, *Conceptual Views on the Activities of the Armed Forces of the Russian Federation in the Information Space*, 2011; MoD, *Defense Minister Sergey Shoygu Called Complete Submission to the West as the Main Goal of the Information War of the West Against Russia*, 2019; MoD, *Military Encyclopedic Dictionary*, undated b; Grudinin, 2011; Nuzhdin, 2000.

enemy] of the ability to resist. The scale and consequences of this are so significant that experts introduced the concept of "information war," in which political, economic, and other goals are achieved through the destruction of the information space of the opposing side and gaining ownership of its information resources. Information weapons have been developed for this purpose, specifically means of electronic warfare, software effects, and others. Information confrontation in war follows two directions: the destruction of information, electronic, [and] computer networks and illegal access to the enemy's information resources (and protection of own information space from the enemy); [and] informational-psychological effects (including those created with technological means) on the population and Armed Forces personnel of the opposing sides.[10]

The Russian military-scientific literature generally reflects this official understanding of information confrontation. Russian military scholars emphasize that IPb is a confrontation between states in the information sphere. V. F. Lata and colleagues broadly define IPb as the "state of relations between countries in the information sphere."[11] A. Nogovitzin similarly defines IPb as a "confrontation between states in the information space with the aim of causing damage to information systems, processes, and resources, critical structures, [and] undermining political and social systems" in order to "destabilize society and the adversary state as a whole."[12] IPb can be used to achieve a wide variety of goals. According to K. I. Sayfetdinov, for example, information confrontation is the "purposeful use of information to achieve political, economic, military, and other goals."[13] The military-scientific literature also emphasizes the

[10] Ministry of Defense of the Russian Federation, undated c.

[11] V. F. Lata, V. A. Annenkov, and V. F. Moiseev, "Informatsionnoye Protivoborstvo: Sistema Terminov i Opredeleniy [Information Confrontation: System of Terms and Definitions]," *Vestnik Akademii Voennykh Nauk [Journal of the Academy of Military Sciences]*, No. 2, 2019, p. 133.

[12] A. Nogovitzin, "Informatsionnaia Voyna: Novii Vizov Budushchego [Information War: A New Future Challenge]," *Armeiskii Sbornik [Army Digest]*, No. 4, April 2009, p. 12.

[13] K. I. Sayfetdinov, "Informatsionnoye Protivoborstvo v Voyennoy Sfere [Information Confrontation in the Military Sphere]," *Voennaya Mysl' [Military Thought]*, No. 7, 2014, p. 38.

distinction between IPb and information warfare. V. Slipchenko characterizes IPb as a broader concept, signifying a "multifaceted, multifactorial"[14] struggle that encompasses "social systems, classes, nations, [and] states through diplomatic, political, informational, psychological, financial, economic influence, armed conflict, and many other forms . . . to achieve strategic and political goals,"[15] while information warfare, according to Lata and colleagues, consists of information operations during active conflict and "emphasizes the specificity of information confrontation in the preparation for and conduct of military (combat) actions."[16] Sayfetdinov emphasizes, however, that IPb is not limited to peacetime and periods of competition. Rather, IPb continues to play an important role during wartime through "gain[ing] and maintain[ing] information superiority over the enemy's armed forces" while simultaneously "creat[ing] favorable conditions for the preparation and use of [Russia's] armed forces."[17] The "essence of information confrontation," according to P. I. Antonovich, "lies in the mutual dependence (vulnerability) of potential adversaries on information and information systems."[18]

Subtypes of Information Confrontation

Information confrontation can be conducted using a wide range of tools, including psychological operations, the provision of moral and psychological support, offensive and defensive uses of information

---

[14] V. Slipchenko, "Informatsionnyy Resurs i Informatsionnoye Protivoborstvo [Information Resources and Information Confrontation]," *Armeiskii Sbornik [Army Digest]*, No. 10, 2013, p. 54.

[15] Slipchenko, 2013, p. 53.

[16] Lata, Annenkov, and Moiseev, 2019, p. 134.

[17] Sayfetdinov, 2014, p. 39.

[18] P. I. Antonovich, "Izmeneniye Vzgliadov na Informatsionnoye Protivoborstvo na Sovremennom Etape [Changing Views of Information Confrontation in the Modern Era]," *Vestnik Akademii Voennykh Nauk [Journal of the Academy of Military Sciences]*, No. 1, 2011, p. 44.

technologies (including both hardware and software), intelligence, and electronic warfare.[19]

In recognition of the diversity of tools used to carry out information confrontation, Russian military scholars have identified two main subtypes of IPb: informational-psychological and informational-technical.[20] The informational-psychological subtype of IPb includes efforts to influence the enemy's population and military forces,[21] including by "mislead[ing] the enemy, undermin[ing] its will to resist, produc[ing] panic in its ranks, and generat[ing] betrayal."[22] Informational-psychological confrontation can be both offensive and defensive; according to V. Ryabchuk and V. Nichipor, it is "aimed at the enemy's thoughts and the defense of one's own thoughts from the same effect from the enemy."[23] Military personnel "not only actively participate in information confrontation," V. Karpuchin explains, but are "themselves an object of continuous informational-psychological influence."[24] It offers a way to "control the

---

[19] Sayfetdinov, 2014, p. 41. According to A. N. Limno and M. F. Krysanov, for example, IPb includes "disguise [and] concealment, psychological operations, intelligence, radio-electronic struggle and programmatic-mathematical influence." A. N. Limno and M. F. Krysanov, "Informatsionnoye Protivoborstvo i Maskirovka Voysk [Information Confrontation and Concealment of Forces]," *Voennaya Mysl' [Military Thought]*, No. 5, 2003, pp. 70–74. S. A. Modestov also characterizes "intelligence and counterintelligence, electronic warfare, camouflage, [and] psychological operations" as tools of IPb. S. A. Modestov, "Strategicheskoye Sderzhivaniye na Teatre Informatsionnogo Protivoborstva [Strategic Containment in the Theater of Information Confrontation]," *Vestnik Akademii Voennykh Nauk [Journal of the Academy of Military Sciences]*, No. 1, 2009, p. 34.

[20] See, for example, Ministry of the Interior of the Russian Federation, "Sredstva i Sposobi Informatsionnogo Vozdeystviya v Sovremennom Mire [Means and Methods of Information Impact in the Modern World]," Saint Petersburg University, Department of Special Information Technologies, April 30, 2020.

[21] K. A. Trotsenko, "Informatsionnoye Protivoborstvo v Operativno-Takticheskom Zvene Upravleniya [Information Confrontation on the Operational-Tactical Level]," *Voennaya Mysl' [Military Thought]*, No. 8, 2016, p. 20.

[22] Sayfetdinov, 2014, p. 38.

[23] V. Ryabchuk and V. Nichipor, "Prognozirovaniye i Predvideniye v Sisteme Planirovaniya Operatsii i Obshchevoyskovogo Boya [Forecasting and Prediction in Operational Planning Systems and Combined Arms Combat]," *Armeiskii Sbornik [Army Digest]*, No. 10, October 2012, p. 38.

[24] V. Karpuchin, "Informatsionnaya Bezopasnost' Voysk [Information Security of Troops]," *Boevaya Vakhta [Battle Watch]*, No. 23, March 28, 2007.

enemy's mind," either directly or indirectly, by introducing specific information, "on the basis of which [the adversary] makes a decision."[25] By "influencing the public consciousness of the population," informational-psychological confrontation "forc[es] the population of the victimized country to support the aggressor, acting contrary to their interests."[26] Some scholars state explicitly that the aim of informational-psychological confrontation is to effect regime change by achieving a "massive influence on the military-political leadership of the adversary."[27] Informational-psychological confrontation can also be used to prolong internal deliberations on policy decisions within the adversary state.[28]

The informational-technical subtype of IPb, on the other hand, involves the physical manipulation of information networks and tools, including the "destruction of information, radio-electronic, [and] computer networks, and [gaining] unauthorized access to the information resources of the enemy."[29] Kuleshov and colleagues note that informational-technical confrontation seeks to influence "communication networks and information networks used by government organizations in the performance of their management functions," "military information infrastructure," "information and management structures of transportation and industrial enterprises, and mass media."[30] It consists of exerting a "software-technical impact on the adversary's information resources," and hardening one's own information resources to prevent such an impact.[31] While recent developments in information technologies have made the impact of informational-technical confrontation

---

[25] B. V. Khudoleev, "Informatsionnoye Protivoborstvo. Kogda Streliayut Slovom [Information Confrontation. When Shooting with Words]," *Krasnaya Zvezda [Red Star]*, No. 193, November 19, 2005.

[26] Y. E. Kuleshov, V. V. Zhutdiev, and D. A. Fedorov, "Informatsionno-Psikhologicheskoye Protivoborstvo v Sovremennykh Usloviyakh: Teoriya i Praktika [Information-Psychological Confrontation in Modern Conditions: Theory and Practice]," *Vestnik Akademii Voennykh Nauk [Journal of the Academy of Military Sciences]*, No. 1, 2014, p. 106.

[27] Kuleshov, Zhutdiev, and Fedorov, 2014, p. 106.

[28] Kuleshov, Zhutdiev, and Fedorov, 2014, p. 106.

[29] Trotsenko, 2016, p. 20; Orlansky, 2008, p. 66.

[30] Kuleshov, Zhutdiev, and Fedorov, 2014, p. 105.

[31] Modestov, 2009, p. 2.

more pronounced, scholars have observed the outsized influence of informational-technical tools for several decades, with one commentator noting in 2001 that because information networks play such a significant role in all aspects of modern militaries, informational-technical confrontation "blurs the line between war and peace" by ensuring that militaries are "in constant information confrontation."[32]

In general, informational-psychological effects target the society of the state under attack, while software effects and the physical destruction of information systems achieve specific effects against informational-technical systems. There is some overlap, however, between these subtypes. S. A. Modestov and colleagues note that social media can be the target of informational-technical confrontation, while also providing "qualitatively new tools for the implementation of propaganda, agitation, and the application of informational-psychological impact on the population."[33] Russian military scholars have emphasized that an effective system of IPb should incorporate both subtypes, including tools such as "concealment (the nucleus of the system), psychological operations, reconnaissance, electronic warfare, and software and mathematical effects."[34] They have also noted the potential broader, long-term impact of IPb as a means of "cultural" seizure in a struggle for cultural superiority in other countries, based on the consideration that the borders of influence of a country are not determined by its physical borders but "by subtle cultural symbols that indicate that we belong to the same world."[35] The features of these subtypes are described in Table 1.3.

---

[32] "Osnovniye Napravleniya Obespecheniya Informatsionnoy Bezopasnosti v Deyatel'nosti Voysk (Sil) [Main Trends in the Information Security of Troops (Forces)]," *Boevaya Vakhta [Battle Watch]*, No. 99, December 22, 2001.

[33] S. A. Modestov, D. A. Nikitin, and E. A. Rabchevsky, "Sotsial'niye Seti kak Teatr Informatsionnogo Protivoborstva v Usloviyakh Sovremennoy 'Gibridnoy' Voyni [Social Networks as a Theater of Information Confrontation in Today's Hybrid War]," *Vestnik Akademii Voennykh Nauk [Journal of the Academy of Military Sciences]*, No. 3, 2019, p. 20.

[34] Limno and Krysanov, 2003, p. 74.

[35] Zh. K. Kenispaev and N. S. Serova, "Civilizatsionnye Voyni: Antichnost [Civilization Wars: Antiquity]," in *Information Wars as a Struggle Between Geopolitical Opponents Civilizations and Ethos*. Collection of Works of All-Russian Scientific Conference, Novosibirsk, April 26–27, 2018.

Table 1.3
Information Confrontation Subtypes

	Informational-Psychological	Informational-Technical
Main tools	• Psychological operations • Disguise, concealment, and camouflage • Provision of moral and psychological support • Intelligence and counterintelligence	• Offensive and defensive uses of hardware and software • Electronic warfare • Disguise, concealment, and camouflage • Reconnaissance • Intelligence and counterintelligence

Main targets of influence	• The military-political leadership of the adversary and their thought processes
	• Adversary military personnel and their thought processes, morale, and motivation
	• Influencing the adversary's intentions, doctrine, tactics, methods of confrontation, conceptions of morality, cohesion of units, level of training and experience, and understanding of the situation
	• The civilian population and their thought processes and consciousness
		• Information and communications networks of government organizations, including the military's information infrastructure
		  — Networks required for the creation, processing, and storage of information
		  — Command and control systems
		  — Communication networks
		  — Intelligence systems
		• Information infrastructure of transport and industrial enterprises, mass media, and social media

Selected activities	• Identifying and suppressing activities that promote harmful ideologies and religious teachings • Dissemination of propaganda, agitation, and seeking psychological effects through social media • Stimulating the activities of civil society organizations and citizens to counter hostile ideologies and religious teachings • Identifying threats and sources of the spread of misinformation about the state and its policies, and reducing the severity of expected negative consequences	• Destroying information, radio-electronic, and computer networks • Gaining unauthorized access to the adversary's information resources • Hardening one's own informational-technical resources to prevent exploitation by the adversary — Neutralizing, eliminating, or reducing the danger of negative consequences of any impact on one's own information infrastructure by hostile (unfriendly) states

SOURCES: Streltzov, 2013; Ministry of the Interior of the Russian Federation, 2020.

Although this terminology is widely used in the Russian military-scientific literature, several scholars have proposed alternative terminologies. Streltzov, for example, explains that IPb can be carried out in two ways: without the use of technological means in the information space (i.e., via confrontation in the area of political ideologies, or political IPb) and with the use of technological means (i.e., via confrontation in the area of information technologies, or technological IPb).[36] According to A. A. Streltzov, political IPb occurs when illegitimate political forces aim to use their influence to infringe upon the political independence of legitimate state actors by undermining public support for the latter. Technological IPb, on the other hand, occurs when one side employs information and communication technologies (or information weapons) to disrupt the stable and secure operation of the opposing side's information infrastructure.[37] Y. E. Kuleshov and colleagues states that IPb consists of three main components: information support, information countermeasures, and information defense.[38]

Despite the perceived importance of information confrontation in Russian strategy, a "unified system of terms, concepts, and definitions" related to IPb remains elusive.[39] As Lata and colleagues note, "a military lexicon of . . . new terms and concepts" related to IPb has arisen in recent years, but every scholar "gives their own definition for each of these concepts," which "significantly complicates the development of a unified understanding of the problem as a whole and its components."[40]

Figure 1.1 below summarizes the basic features of IPb, as discussed in this chapter. It is intended to aid the reader in understanding

---

[36] A. A. Streltzov, "Osnovniye Zadachi Gosudarstvennoy Politiki v Oblasti Informatsionnogo Protivoborstva [Primary Issues for Government Policies in the Area of Information Confrontation]," *Voennaya Mysl' [Military Thought]*, 2013, pp. 18–25. See also A. Raskin and I. Tarasov, "Informatsionnoye Protivoborstvo v Sovremennoy Voyne [Information Confrontation in Modern Warfare]," *Informatsionniye Voyny [Information Wars Journal]*, Vol. 4, No. 32, 2014, pp. 2–6.

[37] Streltzov, 2013, pp. 20, 22.

[38] Kuleshov, Zhutdiev, and Fedorov, 2014, p. 105.

[39] Lata, Annenkov, and Moiseev, 2019, p. 129.

[40] Lata, Annenkov, and Moiseev, 2019, p. 129.

Figure 1.1
Basic View of Information Confrontation and Its Components

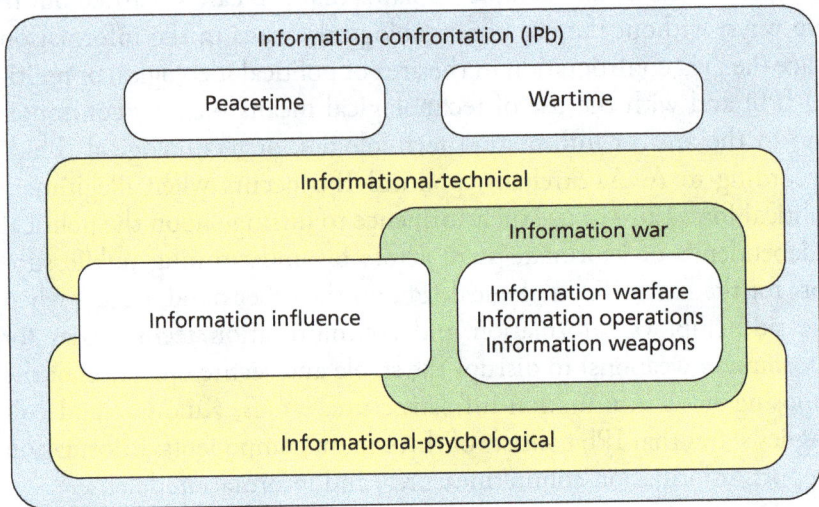

Information confrontation (IPb)

Peacetime

Wartime

Informational-technical

Information war

Information influence

Information warfare
Information operations
Information weapons

Informational-psychological

the more detailed discussions of the various components of IPb and related concepts and terms that will be introduced in subsequent chapters.

Entities Involved in Information Confrontation

The two main groups responsible for executing information confrontation can be separated into state and nonstate actors. State actors include military organizations, security services, and other federal agencies. Nonstate actors include "patriots," ideological extremists, and terrorists. Table 1.4 summarizes the primary state entities that are involved in information confrontation, while Table 1.5 summarizes the primary nonstate entities that are involved in information confrontation.

Figure 1.2 depicts the general relationship of the military organizations (listed in Table 1.4) within the armed forces of the Russian Federation.

Table 1.4
State Entities Involved in Information Confrontation

	Entity	Role in Information Confrontation
Military organizations— operations	Glavnoye Razvedyvatel'noye Upravleniye (GRU [Main Intelligence Directorate])	• Russian military intelligence agency that carries out hacking activities aimed at "areas of concern and political tension" • Operatives were involved in cyberattacks on computers of the Democratic Party figures prior to 2016 U.S. presidential campaign • Oversees troops that work with several "research companies," whose activities may include those of the "Fancy Bear" group, cryptography, and hacking assignments (or APT 28)
	"Information Operations Forces"	Troops within the Russian armed forces • Focused on cybersecurity operations, specifically defending against adversary computer network operations during military conflict • Responsible for psychological operations
	8th Directorate of the General Staff of the Russian Federation	• Works with mass media, social media, and foreign militaries to protect the Russian military's operational security
	Voyska Radioelektronnoy Bor'by (REB Troops [Electronic Warfare Troops])	• Russian military's electronic warfare branch • Main role is "winning and retaining superiority in command and control of combat actions"
Military organizations— training and education	Military universities	• Prepare some cadets to carry out defense of information • For example, graduates of Krasnodar General S. Shtemenko Institute go on to serve at the 8th Directorate of the General Staff of the Russian Federation
	Research companies	• New military conscripts receive training to work in information security, cryptography, information defense, and countering technical intelligence and may serve a one-year tour at Krasnodar Institute • The 6th Research Company is known as the "military cyber-defenders" • Some research companies are subordinate to the GRU

Table 1.4—Continued

	Entity	Role in Information Confrontation
Federal agencies	Federal'naya Sluzhba Bezopasnosti (FSB [Federal Security Service]), FSB Institute	• Involved in Russian hacking activities aimed at "areas of concern and political tension" • May be affiliated with APT 29 • Believed to recruit individuals involved in cybercrime
	Sluzhba Vneshney Razvedki (SVR [Foreign Intelligence Service])	• Possibly involved in hacking activities aimed at "areas of concern and political tension"
Individuals (e.g., "career hackers")		• Some individuals may conduct hacking activity on behalf of the Russian military and security services (e.g., the Internet Research Agency,[a] "paid civil trolls") • The Kremlin has also pioneered the use of "bots," or fake social media accounts, that are fully or partially automated or operated anonymously by humans; trolls complement the use of bots. • Some individuals may also be conscripted into Russia's armed forces
State-run media		• Used for propaganda purposes and disinformation campaigns, to influence political views of the general public
State elites		• State elites use social networking platforms, and other communication resources, to influence and manipulate public consciousness, often for their own benefit

SOURCES: RAND analysis of select Russian military-scientific literature (Lysenko and Brooks, 2018; Radio Free Europe/Radio Liberty, 2018; Rossiyskiy Ekonomicheskiy Universitet Imeni G. V. Plekhanova [Russian Economic University], website, undated; "Disinformation Report on Foreign Interference in the 2016 Election," Yonder website, 2018; Palitay, 2018; Podvigin, 2018; Muhin and Rekunko, 2017; Fraher and Arkhipov, 2017; Filipenko, 2017; Giles, 2011; Helmus et al., 2018).

[a] The Internet Research Agency LLC is a troll farm located in Saint Petersburg, Russia.

The Main Directorate of the General Staff resides at the top of this structure. The Information Operations Forces consist of personnel in the Special Services Centers who specialize in *Zarubezhnaya Voennaya Informatsiya/Kommunikatsiia* (ZVIK) [Foreign Military Information and Communication], psychological operations (PsO), and operations

Figure 1.2
Military Organizations Force Structure

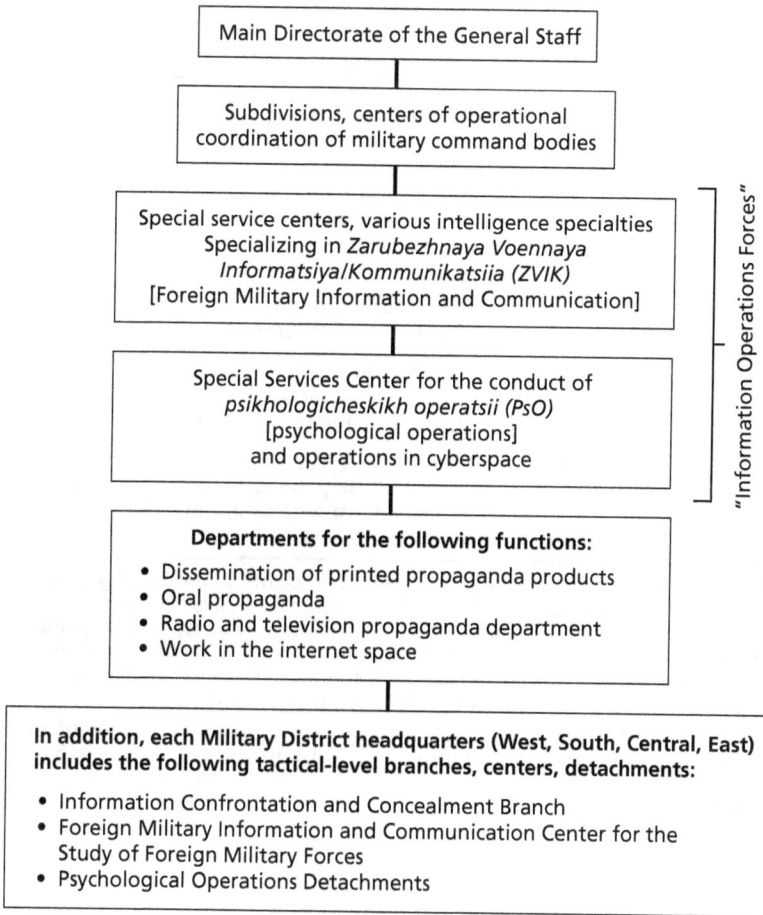

```
              Main Directorate of the General Staff
                            |
              Subdivisions, centers of operational
              coordination of military command bodies
                            |
          Special service centers, various intelligence specialties
               Specializing in Zarubezhnaya Voennaya
                 Informatsiya/Kommunikatsiia (ZVIK)
             [Foreign Military Information and Communication]
                            |
                Special Services Center for the conduct of
                    psikhologicheskikh operatsii (PsO)
                        [psychological operations]
                      and operations in cyberspace
                            |
              Departments for the following functions:
             • Dissemination of printed propaganda products
             • Oral propaganda
             • Radio and television propaganda department
             • Work in the internet space
                            |
  In addition, each Military District headquarters (West, South, Central, East)
  includes the following tactical-level branches, centers, detachments:
  • Information Confrontation and Concealment Branch
  • Foreign Military Information and Communication Center for the
    Study of Foreign Military Forces
  • Psychological Operations Detachments
```

"Information Operations Forces"

SOURCES: RAND analysis of select Russian sources (Filipenko, 2017; MoD website, n.d.).

in cyberspace. PsO forces are subordinate to the forces specializing in ZVIK. The PsO Center has several subordinate departments that perform functions associated with propaganda and work conducted using the internet. At the tactical level, each military district includes entities dedicated to information confrontation, ZVIK, and PsO.

Table 1.5
Nonstate Entities Involved in Information Confrontation

Entity	Role in Information Confrontation
Universities	Some universities contract with the Russian Ministry of Defense to provide services, such as software development, for the Russian armed forces (e.g., Saint Petersburg ITMO University)
Private businesses (e.g., social media companies,[a] smartphone companies)	Viewed as sources of large volumes of publicly available information, as well as "agents" of "modern spying" who are "doing the listening" (e.g., using smartphones)
Media (not state-run)[a]	• Source of pro-Western propaganda • Carry out disinformation against Russia and Russian interests abroad
Terrorist organizations[a]	• Work to destabilize Russian domestic politics • Spread ideological extremism that is damaging to Russian domestic politics
Private citizens ("patriots")	"Patriotic hackers" or "'unpaid' cyber-patrol 'volunteers'" may conduct hacking activities on their own in response to stories in the foreign press that appear to "malign" their "homeland"

SOURCES: RAND analysis of select Russian military-scientific literature (Lysenko and Brooks, 2018; Radio Free Europe/Radio Liberty, 2018; Rossiyskiy Ekonomicheskiy Universitet Imeni G. V. Plekhanova [Russian Economic University], website, undated; "Disinformation Report on Foreign Interference in the 2016 Election," Yonder website, 2018; Palitay, 2018; Podvigin, 2018; Muhin and Rekunko, 2017; Fraher and Arkhipov, 2017; Filipenko, 2017; Giles, 2011; Helmus et al., 2018).

[a] While some entities conduct information confrontation *for or on behalf of* Russia, others are viewed as conducting information confrontation *against* Russia. Social media companies, nonstate affiliated media, and terrorist organizations are among the latter.

CHAPTER TWO

Information Confrontation in Historical Perspective

The concept of information confrontation has deep roots in Russian (and Soviet) military thinking. The Russian military-scientific literature characterizes the use of information as a constant in "all stages of historical development."[1] Information confrontation is a phenomenon that "originated in ancient times," arising "simultaneously with the emergence of armed confrontation as an integral part of the armed struggle."[2] It is noted in the Russian military-scientific literature that the modern concept of IPb is the result of four previous stages of historical development, each of which is based on the emergence of new technologies, "verbal, paper, technical, and telecommunications," with "each subsequent stage absorb[ing] the means and methods of the previous stage and develop[ing] them."[3]

This chapter traces the historical antecedents of the modern concept of IPb, from the late 18th century to the post–Cold War era.

---

[1] Lata, Annenkov, and Moiseev, 2019, p. 128.

[2] Kuleshov, Zhatdiev, and Fedorov, 2014, p. 104.

[3] B. M. Gryzlov and A. B. Pertsev, "Informatsionnoye Protivoborstvo: Istoriya i Sovremennost [Information Confrontation: History and Modernity]," *Vestnik Akademii Voennykh Nauk [Journal of the Academy of Military Sciences]*, No. 2, 2015, p. 124.

Imperial Russia

Psychological warfare[4] and information operations, often referred to as "spetsprop," are elements of information confrontation with a long history in Russian and Soviet military strategy. The Russian military has a centuries-long history of applying asymmetric warfare,[5] including spetsprop, to opponents' weaknesses. Early psychological operations, often facilitated by handwritten leaflets and rumormongering behind enemy lines, aimed to destabilize enemy ranks and coalitions by reducing their morale and will to fight Russian troops. In 1799, General Suvorov, one of Russia's most renowned military heroes, disseminated messages to Piedmontese soldiers that persuaded many to defect to the Russo-Austrian army, representing one of the earliest cases of Russian military psychological warfare.[6]

The use of psychological warfare during the Napoleonic Wars of the early 19th century played a significant role in shaping later Russian military thought on psychological operations as a means of reducing the morale of enemy forces and influencing their decisionmaking. While conducting raids behind enemy lines, Russian light cavalry and partisans helped disseminate leaflets addressed to locals, primarily asking

---

4   Russian strategic documents do not define this term, which is also absent in the Russian *Military Encyclopedic Dictionary*. The dictionary includes a related term, *psikhologicheskaya bor'ba* (psychological struggle), defined as activities of adversaries in the informational-psychological sphere that are aimed at changing the behavior and attitudes of certain groups of people in a desired direction; these activities take place during both the peacetime and conflict. Ministry of Defense of the Russian Federation, "Psikhologicheskaya Bor'ba [Psychological Struggle]," *Voyennyy Entsiklopedicheskiy Slovar' [Military Encyclopedic Dictionary]*, undated e.

5   The Defense Terminology Depository of the Russian Ministry of Defense includes a closely related term, *asimmetrichnyye voyennyye (boyevyye) deystviya* (asymmetric military [combat] activities), defined as "military (combat) activities, characterized by inequality of forces, means, technologies, and resources of the warring parties." These activities include guerrilla, reconnaissance, sabotage, and terrorist activities. Ministry of Defense of the Russian Federation, "Asimmetrichnyye Voyennyye (Boyevyye) Deystviya [Asymmetric Military (Combat) Activities]," in *Spravochnik Po Terminologii v Oboronnoy Sfere [Defense Terminology Repository]*, undated a.

6   L. V. Vorontsova and D. B. Frolov, *Istoriya i Sovremennost' Informatsionnovo Protivoborstva [History and Modernity of Information Confrontation]*, Goryachaya liniya-Telekom, 2006, p. 22.

them to help resist, and enemy forces, calling on them to surrender or attempting to sow divisions among them. In June 1812, Russian units disseminated a message to German soldiers fighting for Napoleon from Russian Minister of War Barclay de Tolly asking them to question what they were fighting for and to join the "German legion" fighting alongside Russia against a regime that oppressed their nation.[7] Russian military propaganda became more prominent as Napoleon drove further into Russian territory and his forces became stalled in a difficult campaign. Leaflets such as de Tolly's were seen as both necessary and— as conditions for Napoleon's army became worse—increasingly effective. German units, for example, were six times more likely to surrender than their French counterparts during the fall of 1812.[8] Russian military personnel often used face-to-face communications to disseminate propaganda directly to their targets: either during frontline negotiations or occasionally by disguising as the enemy, entering their camp, and spreading demoralizing information to enemy forces.[9]

The Great War and the Revolution

During World War I, the Russian military made increasing use of propaganda, though its scale was seldom significant enough to affect any change in enemy morale or decisionmaking. At the onset of the conflict, the Ministry of Foreign Affairs was responsible for external political propaganda, exclusively under the Department of Printing and Informing. However, Imperial military culture prevented the General Staff from using disinformation, lies, and slander in propaganda, which reduced the effectiveness of messaging at the front as well as strategic campaigns in neutral countries.[10] As a result, the military and Ministry

---

[7] R. E. Al'tshuller and A. G. Tartakovskiy, *Listovki Otechestvennoy Voyny 1812 Goda [Leaflets of the Patriotic War of 1812]*, U.S.S.R. Academy of Sciences, 1962, pp. 23–24.

[8] S. I. Repko, "Voyna i Propaganda [War and Propaganda]," *Novosti [The News]*, 1999, p. 85.

[9] Repko, 1999, p. 110.

[10] Repko, 1999, pp. 144–146.

of Foreign Affairs frequently criticized this department for being ineffective.[11] Despite these hurdles, military propaganda was employed in a broader sense than ever before. Between 1914 and 1917, Russia's military used leaflets and print products to entice enemy troops to surrender, incite non-German and non-Austrian ethnic groups to revolt, and promote intervention from neutral countries through popular and private channels. Slavic nationalities behind enemy lines were especially targeted, such as a 1916 leaflet disseminated to the Austro-Hungarian 9th Army calling on Croatians to turn their weapons against their oppressors:

> Austria has always only turned its attention to you when it is in need and always when it requires blood . . . after the war, Austria will be more wicked than before and will want to erase from the face of the earth your brave and noble race. Your great mother-Russia calls you to join it, here and now the center of all Slavic peoples.[12]

Throughout World War I, the General Staff gained experience in strategic propaganda, including that aimed at the United States. Through ostensible third-party press outlets in New York, Chicago, and San Francisco and the use of communist front organizations, the General Staff controlled unattributable messaging designed to influence public opinion in the United States, such as bulletins provided to American agents of the Associated Press and United Press that were distributed to over 1,850 foreign newspapers and journals.[13]

The Bolshevik Red Army immediately employed asymmetric tactics in the civil war and postrevolution conflict. By using psychological operations, the Red Army between 1917 and 1920 was able to cleave Terek and Kuban Cossacks from the army of former tsarist general Anton Denikin by broadcasting guarantees about land ownership, sowing discord among Cossack ranks through special agents, and

---

[11] A. B. Astashov, *Propaganda na Russkom Fronte v Gody Pervoy Mirovoy Voyny [Propaganda on the Russian Front During the First World War]*, Spetskniga, 2012, p. 11.

[12] Astashov, 2012, p. 290.

[13] Repko, 1999, p. 133.

allowing Cossacks to surrender unharmed.[14] The Bolsheviks enjoyed similar successes against foreign militaries during the civil war. Overt and covert confrontation with Poland, which lasted virtually from the onset of the Soviet–Polish War in 1921 until the annexation of Poland by Nazi Germany and the Soviet Union in 1939, led to the employment of a wide range of asymmetric warfare techniques, including both sides' efforts to exploit ethnic rifts within each other's borders through psychological and intelligence operations.[15] Many early Soviet military leaders first served in the prerevolutionary Bolshevik underground, which provided formative experiences with intelligence operations and propaganda that they would use in their military intelligence careers.[16]

The Interwar Years

The first decade of the Soviet government's existence allowed for a tenuous peace that provided Red Army strategists time enough to reflect on Great War and revolutionary experiences. Soviet intelligence and military intelligence evolved quickly during this period, influenced heavily

---

[14] In August 1918, the early Soviet government established a secret department to use propaganda and intelligence operations to undermine the relationship between the White Army and Cossacks operating in southern Russia, while local party organizations created special unofficial centers to manage agents sent to Cossack formations in order to turn them against the Red Army's enemies. Between August and December 1918, these agents disseminated around 230,000 leaflets to Cossacks fighting alongside the White Army, and the Soviet government afforded these operations some 46,000 rubles, a significant sum for the cash-strapped revolutionary government. Similarly, per Lenin's directive, 150 million rubles were given to Bashkir units fighting for the White Army to instead join the Bolsheviks, which was accompanied by a propaganda effort that included promulgations from Trotsky that promised the Bashkirs self-governance after the revolution, which resulted in their quick defection from pro-tsarist forces. See Repko, 1999.

[15] A. A. Zanovich, "Pol'skaia Razvedka Protiv Krasnoi Armii, 1920–1930-e gody [Polish Intelligence Against the Red Army, 1920–1930s]," *Voenno-Istoricheskii Zhurnal [Military and History Journal]*, No. 10, 2007, pp. 32–36.

[16] The demise of Yan Berzin, the head of Soviet military intelligence, as part of Stalin's purges in the latter 1930s perhaps represents a squandered opportunity to integrate psychological with kinetic special operations and intelligence work, as Berzin, among others extinguished through the Great Terror, were well versed in all of these operations.

by diplomatic isolation and the degree to which Soviet leadership saw threats encircling the borders, from Helsinki to Khabarovsk. During this period, a cultural shift allowed for the inclusion of distortion and fabrications in propaganda. Drawing on their experiences during the revolution and civil war, moreover, the Red Army leaders explicitly characterized psychological operations as playing an important role in special operations. For instance, a 1927 manual on military intelligence published by the Red Army's General Staff emphasized the importance of influencing populations behind enemy lines to participate in rearguard attacks. As two senior staff officers with the Defense Commissariat wrote,

> Political sentiment of the population in an enemy's rear plays a big role in an opponent's successful activities; because of this it is extremely important to generate sentiments among populations against the enemy and use them to organize people's uprisings and partisan detachments in the enemy's rear, which would totally or partially destroy the proper function of rear-echelon work and populated centers of the enemy.[17]

Despite initiatives to rapidly boost the conventional strength of the Red Army, throughout most of the interwar period, Soviet leadership perceived a disparity between their own military capabilities and those of their potential adversaries. Because of this, Soviet propaganda highlighting the technical prowess of the Red Army became one of the most important themes of the 1920s and 1930s. It targeted the rank and file of the Soviet military as well as external military and nonmilitary audiences. A 1921 manual on military propaganda, for example, proposed the use of brochures, leaflets, and journals to familiarize soldiers with the features of modern warfare, including new technology, to make best use of the "breathing space" (*peredyshka*) between the civil war and impending conflict with "imperial powers."[18]

---

[17] K. Shil'bakh and V. Sventsitsii, *Voennye Razvedki [Military Intelligence]*, Voennoe tipografnoe upravlenie, 1927, p. 86.

[18] *Voennaia Propaganda v Armii: Materialy [Military Propaganda in the Army: Materials]*, Khar'kov izdanie ukrpura, 1921.

At the same time, Soviet authorities ordered military intelligence to invigorate disinformation efforts aimed at likely adversaries that inflated the numerical and technological strength of the Red Army. For example, Leonid Trotsky, the early Soviet Commissar for Defense, ordered military intelligence to exaggerate the quantitative and qualitative capacity of the Soviet military by 50 to 60 percent. This led to a years-long campaign in which false information regarding the capacity of the Soviet ground and air forces, as well as the Soviet naval forces, was fed to Western powers.[19] During the "war scares" of the late 1920s, Soviet authorities relied on propaganda to dispel widespread views in the technological superiority of its adversaries. As a collection of materials for military propaganda published by Soviet authorities in Crimea stated, one of the most significant hurdles to local propaganda was the "incorrect, exaggerated" view that foreign militaries' technological capabilities dwarfed those of the Red Army.[20] By the time the Soviet Union openly engaged enemies along its peripheries in the late 1930s, propaganda was disseminated to enemy soldiers that emphasized the futility in fighting against the powerful tanks and aircraft of the Red Army.[21] These efforts had mixed success, as the observed deficiencies and operational shortcomings of the Red Army during the Soviet–Finnish War and campaigns at Khalkhin-Gol and Lake Khasan in the Far East were at odds with the propaganda themes surrounding Soviet military predominance, technical or otherwise.

Conflicts in the Far East and Europe during the late 1930s, military assistance, and other interventions provided a wide range of case studies for Soviet military thought about information operations as a tool of asymmetric warfare. Soviet military authorities understood

---

[19] Evgenii Gorbunov, *Stalin i GRU: 1918–1941 gody [Stalin and the Main Intelligence Directorate: 1918–1941]*, Rodina, 2018, pp. 56–59.

[20] *Materialy dlia Dokladov po Voennoi Propagande [Military Propaganda Reporting Materials]*, Krymgosizdat, 1928.

[21] For instance, frontline propaganda during the campaign at Khalkhin-Gol in 1939 highlighted the superiority of the Soviet air force over Japanese aircraft. *Partiino-Politicheskaia Rabota v Boevoi Obstanovke [Political and Party Work in a Combat Situation]*, Voennizdat, 1940, pp. 121–131.

there to be an increasingly strong link between intelligence and propaganda efforts.[22] Between 1939 and 1940, military authorities developed the "seventh department" of the military's political directorate, responsible for propagandizing enemy and neutral target audiences.[23]

The Great Patriotic War

The political chaos brought about by Stalin's Great Terror in the late 1930s had a deleterious effect on the conceptualization and implementation of information operations. Leading intelligence figures, many of whom had experience in partisan warfare and frontline propaganda, were mostly purged from the ranks of military leadership.[24] It would take the dire circumstances of World War II to relearn lessons and revive capabilities lost through extensive purges.

Over the course of World War II, the quality and effectiveness of Soviet military propaganda gradually improved. At first, messaging aimed at Axis soldiers fell flat, as Soviet leaflets contained Marxist-Leninist jargon unfamiliar to enemy troops and calls for surrender were unpersuasive in light of repeated Soviet military catastrophes as the Wehrmacht drew closer to Moscow in the early months of the German invasion. As the tide started to turn slowly in Moscow's favor, military propaganda began to reinvent itself and become more effective at influencing the enemy. Spetsprop officers, for instance, recognized the pivotal role of German officers in persuading frontline units to surrender to the Red Army, which led spetsprop to reorient from heavily messaging the rank and file, largely with hopes that soldiers would revolt against their "bourgeois" leadership, to German commanders who could often

---

[22] Vitalii Zharkov, *Politicheskaia Rabota v RKKA (1929–1939 gg.) [Political Work in the Red Army (1929–1939)]*, K. D. Ushinskii Yaroslavskii State Pedagogical University, 2005, p. 120.

[23] For a thorough account of the early history of the Red Army's external propaganda efforts, see M. I. Burtsev, *Prozrenie [Epiphany]*, Voenizdat, 1981.

[24] For example, Yan Berzin, considered one of the founders of Soviet military intelligence and who had wide experience in propaganda and intelligence activities that dated back to the Russian Revolution, was purged and executed in mid-1938. Yan Gamarnik, who commanded both the intelligence and political directorates during his career, was also purged, though he committed suicide before receiving official punishment for his ostensible treason.

hand over entire formations to nearby Soviet units. As the Red Army's collection of enemy prisoners grew, the effectiveness of its propaganda correspondingly increased. German, Italian, Hungarian, Romanian, and other prisoners were repurposed into frontline broadcasters and proofreaders for proposed products.[25] Aside from many operational lessons in frontline propaganda, such as its efficacy and importance when facing surrounded enemy units, the war demonstrated yet again the ability to fracture enemy coalitions along national and ethnic fault lines, evidenced by the disproportionate number of prisoners of war among the countries fighting alongside Germany.[26]

The experience of World War II provided important lessons for the postwar Soviet military. Official designation for special and psychological operations during the war, under the GRU and military-political directorate, secured their place in the Soviet order of battle with far more certainty than the prewar period allowed; the same is true for Russian military psychological operations. A 1999 textbook on "psychological warfare," for example, refers to World War II and prewar campaigns, particularly the Red Army's invasion of Poland in 1939, as examples of the effectiveness of psychological operations aimed at national differences in an opposing force.[27]

---

[25] Perhaps the most famous national front run by the Red Army's propagandists during the war was the "Free Germany National Committee," which consisted of German prisoners of war that Soviet political officers used to generate and disseminate propaganda, though the movement also directly participated in combat operations later in the war. Among some of the movement's more notable figures was the great-grandson of German patriarch Otto von Bismarck, a Luftwaffe officer captured by the Soviets during the Stalingrad campaign. Evgenii Torsukov, "Pravnuk Bismarka Sotrudnichal s Krasnoy Armiey [Bismarck's Great-Grandson Collaborated with the Red Army]," *Nezavisimmoe Voennoe Obozrenie [Independent Military Review]*, No. 27, 2003, p. 5.

[26] According to Soviet sources, Austrian and Hungarian soldiers were the least receptive groups to spetsprop messaging, as only 14.9 percent of Austrian prisoners and 9 percent of Hungarians captured by the Soviets elected to sign testimonials offered by Soviet captors, while Romanian soldiers were among the most receptive groups to frontline propaganda, 30 percent of which chose to lay down their arms and collaborate with the Red Army. See Repko, 1999, pp. 414–416.

[27] The same textbook, a GRU manual, according to reporters with the *Moscow Times*, laments the cultural encroachment of the West, particularly the United States, on Russia. Vladimir Gavrilovich Krysko, *Sekrety Psikhologicheskoi Voiny [Secrets of Psychological War]*, Minsk, 1999.

The Cold War

The further development of information operations reflects a constant refrain in Soviet and Russian military thought: the growth of new capabilities is often hindered by internal politics, yet subsequently it is reinvigorated by adversaries' activities.[28] The Soviet military's *spetsprop* remained subordinate to the Main Military-Political Directorate throughout the Cold War, which remained institutionally rooted in wartime experience that it transferred to Cold War–era cadres through education. More than likely, spetsprop played a secondary role in strategic information warfare with the West, with the KGB's notorious "Service A" in the vanguard of the covert ideological struggle to weaken adversaries and gain support elsewhere. Like their kinetically oriented counterparts, spetsprop officers nonetheless took part in campaigns abroad to install pro-Soviet regimes around the world and solidify their support. As a Soviet adviser to North Vietnam described, these experiences profoundly affected spetsprop officers serving at the time, particularly the apparent ability of outgunned, small militaries to undermine predominant foes through psychological warfare:

> For me personally, the moment of "scientific truth" came after a trip to Vietnam in 1969, which at that time was heroically fighting against U.S. aggression. Smelling gunpowder during two weeks of being under American bombs in Hanoi, in Haiphong, and at the positions of Vietnamese rocketeers, holding Vietnamese leaflets addressed to American soldiers in my hands, I seemed to sober up and looked at my duties more responsibly as . . . a specialist on the U.S. military. The problem of morale in the experience of the U.S. war against Vietnam demanded immediate scientific reflection.[29]

---

[28] Another clear example of this phenomenon is the early development of spetsprop in the Red Army. Though Stalin and his loyalists did much to extinguish political officers with foreign specialization from the ranks, the fact that the Japanese army at Khalkhin-Gol began propagandizing local Soviet forces drove Moscow to reciprocate. See Burtsev, 1981.

[29] Yurii Petrovich Gusev, "Spetsial'naia Propaganda v Moei Zhizni [Special Propaganda in My Life]," undated.

One commentator similarly noted the importance of informational-psychological confrontation during the Vietnam War, writing that the North Vietnamese had "learned the principles of informational-psychological confrontation and successfully applied" these principles against American soldiers in Vietnam.[30]

At the same time, spetsprop officers wrestled with rapid technological changes in mass communication, which they saw as increasingly important to Cold War competition. A 1986 textbook on propaganda emphasized the growing role of television and—apparently more important—radio, which had become the most important medium to reach international audiences during the "dramatic changes" of the 1970s.[31] Spetsprop and political officers in the Soviet military similarly recognized the increasing potency of television as it applied to military propaganda. A 1983 manual on the "technical means" of propaganda, for instance, noted that television broadcasts would transcend "state borders" in the near future.[32]

The Soviet–Afghan War, which began in 1979 and ended in Soviet withdrawal a decade later, was a formative operational experience for spetsprop. As the personal account of one spetsprop officer shows, the war brought spetsprop closer to intelligence operations as well as more extensive experience in working with partner forces and conducting civil affairs:

> [A] definite breakthrough in the organization of special propaganda in Afghanistan was the creation of the first multipurpose combat agitation and propaganda task-forces [*vneshtatnie otrady*]. . . . Each

---

[30] Y. Kozhuk, "Zametki Voyennogo Obozrevatelya: Novoye Staroye Oruzhiye [Notes of a Military Observer: New Old Weapons]," *Flag Rodiny [Flag of the Motherland]*, No. 12, 2002b.

[31] P. S. Gurevich, *Propaganda v Ideologicheskoy Bor'be [Propaganda in Ideological Confrontation]*, Vyshaya shkola, 1987, p. 160.

[32] V. I. Kondiurin and E. G. Tiutiunik, *Tekhnicheskie Sredstva Propagandy [Technical Means of Propaganda]*, Voennizdat, 1983, p. 6. *Spetsprop* was a subset of broader active measures, or *aktivnyye meropriyatiya*, which were carried out throughout the Cold War. Active measures encompassed a wide range of activities, including disinformation, forgeries, the use of front groups, and broadcasting. Active measures, which were used as an offensive instrument of Soviet foreign policy and were generally covert, differ from IPb, which can be either offensive or defensive and include overt activities.

detachment had an armored and lightweight loud-speaking sta-tion, a field auto club [*pokhodniy autoklub*], a medical vehicle, and military security. As a rule, Afghan officers, local leaders, mullahs, and concert groups worked in [them]. Often scouts, engineers, and an adviser were connected to them, and they could carry out com-prehensive propaganda events while providing all kinds of assis-tance: from medical and food to mine clearance and well drilling.[33]

Transition to the Post–Cold War Era

Although the Cold War served as a formative period in the develop-ment of psychological warfare and information operations, the term "information confrontation" was not present in either Western or Soviet lexicon throughout most of this period. Instead, as is clear from the preceding discussion, terms like "propaganda," "disinformation," "active measures," "reflexive control," and other related concepts were common parlance among Soviet and Western civilian and military cir-cles throughout the Cold War.

The term "information confrontation" appears to have made its debut in the Soviet military-scientific literature in the tumultuous final years of the Cold War.[34] A 1990 article identifying priority areas for analysis penned by then–Lieutenant General I. S. Lyutov marked the first appearance of "information confrontation" in Soviet/Russian military-scientific scholarship, although Lyutov did not explic-itly define the concept.[35] Even so, Lyutov cited the concept within a broader commentary on the growing informatization of the military field. Lyutov perceived that the informatization of society would have

---

[33] Vladimir Ramus' "Osobiy Front [Special Front]," *Suvorovskii Natisk [Suvorov Onslaught]*, No. 16, 2019, p. 6.

[34] Though this instance does not technically fall within the post–Cold War period, we include it here because its emergence fell so close to the collapse of the Soviet Union, and because it appears to have only gained prominence in post-Soviet scholarship.

[35] I. S. Lyutov, "Voennaya Nauka: Prioritetnye Napravleniia Razvitiia [Military Science: Priority Areas of Development]," *Informacionnyj Sbornik Shtaba Ob'edinennyh Vooruzhen-nyh Sil Gosudarstv-Uchastnikov Varshavskogo Dogovora [Information Digest of the Joint Armed Forces Headquarters of the State Parties to the Warsaw Pact]*, June 30, 1990.

profound effects for the military, especially command and control (C2) systems and structures. Though radio- and electronic-enabled C2 systems expedited information processing and decisionmaking, they also introduced major vulnerabilities. A military's susceptibility to electronic warfare attacks would not only cost troops individual victories but would also severely undermine that military's overall potential.[36]

As Lyutov saw it, in modern war, the combatant with superior information collection, decisionmaking, and C2 systems would have the upper hand. Information confrontation, in this context, was characterized as a competition between states to collect more detailed intelligence, to inform a superior operational picture, to support better decisionmaking, and to ensure speedier command and control of troops through the employment of sophisticated information technologies. Though he did not explicitly name it as such, Lyutov described what would later be christened "information superiority."

Just three months after Lyutov's article was published, A. Ya. Vayner, a retired Soviet colonel, authored an article echoing similar themes.[37] Vayner acknowledged that Soviet military science had long considered effective C2 a determinant of successful military campaigns. He noted, however, that unlike before, the explicit paralysis of an adversary's C2 systems and processes was taking on an increasingly prominent role in the conduct of warfare. He characterized confrontation in the informational sphere as a persistent struggle between adversaries for information and intellectual superiority, which he argued could materialize in several ways. They could be classified along two general axes. First, efforts could be offensive and/or defensive—a distinction that Lyutov had also made. Second, efforts could be technical and/or intellectual in nature. Here, Vayner expanded on Lyutov's conception of IPb, which had primarily focused on impacts to tangible C2 systems. Vayner used the term intellectual

---

[36] Lyutov, 1990.

[37] A. Ya. Vayner, "O Protivoborstve v Sfere Upravlenia [On Confrontation in the Sphere of Command and Control]," *Voennaya Mysl' [Military Thought]*, No. 9, 1990.

confrontation to denote activities aimed at influencing an adversary's attitudes and behaviors (offensive) or activities to insulate one's own ranks from such attacks (defensive), similar to the contemporary characterization of informational-psychological confrontation. If, for instance, a state sought to influence the decisions made by adversary leaders, it could corrupt the adversary's data or replace factual information with disinformation. It could also achieve information superiority by selecting intellectually robust commanders, training personnel, and enabling better decisionmaking through the application of technological advancements.

Vayner's threat perceptions and understanding of information confrontation were informed by his observations of North Atlantic Treaty Organization (NATO) rhetoric, exercises, and acquisitions.[38] NATO leadership, according to Vayner, prioritized the automation of information collection, analysis, and dissemination as a means for achieving information superiority. Likewise, a NATO offensive would likely target an adversary's C2 systems. He predicted that an overwhelming electronic warfare strike on adversary C2 systems, the consequences of which could rival a nuclear strike, would serve as NATO's modus operandi at the outbreak of future hostilities.[39]

Lyutov's and Vayner's framing of IPb through a C2 lens was not anomalous for this period. As noted below, a number of early post–Cold War writings discuss IPb in this context. This emphasis is explained by developments in civilian computing, military science, and military theory in the 1970s and 1980s—developments that had been integrated into C2 systems by the early post–Cold War years. Though states have long undertaken efforts to undermine adversaries' chains of command, in the 1970s, the notion of deliberately attacking a competitor's C2 systems through electronic means rose to prominence in both Soviet and U.S. military circles. Prompted by concerns over an emerging Soviet concept known as radio-electronic combat, a number of groups within the Pentagon shifted their gaze to a new

---

[38] Vayner, 1990, pp. 22–23.

[39] Vayner, 1990, pp. 22–23.

U.S. concept in years following Vietnam—C3CM, or command, control, and communications countermeasures.[40] In this period, C3CM became the focal point of a number of prominent Department of Defense (DoD) and Air Force studies designed to shape and mature the concept.

By 1979, the Pentagon's C3CM concept and strategy were formalized in Directive 4600.4. *Command, Control, and Communications Countermeasures*. As conceived of in the directive, C3CM involved the use of friendly capabilities to "influence, degrade, or destroy enemy command, control, and communications" while protecting friendly C2 from such attacks.[41] Subsequent DoD policy, like the 1983 Chairman of the Joint Chiefs of Staff Memorandum of Policy 185, *Command, Control, and Communications Countermeasures*, continued to emphasize the same undercurrents of targeting an adversary's C2 while safeguarding one's own C2 from such assaults. Though the concept was rebranded in 1990 as "command and control warfare," or C2W, its anatomy remained largely unchanged. The one exception was the DoD's addition of psychological operations (psyops) to the list of capabilities that could be leveraged to conduct C2W.[42]

After years of refinement, the United States' integrated C2W concept moved from the pages of doctrine to the battlefield in the Gulf War, as onlookers from the ranks of the Soviet military-scientific community observed from the sidelines:

> [In preparation for] Operation Desert Storm, the United States and its coalition allies were able, for the first time, to bring together the four classic elements of C2W—operations security, military

---

[40] Charles F. Smith, "Command, Control and Communications Countermeasures (C3CM)," *Military Review*, 1983; Norman B. Hutcherson, *Command and Control Warfare: Putting Another Tool in the War-Fighter's Data Base*, Maxwell Air Force Base, Ala: Air University Press, September 1994, p. 2.

[41] Department of Defense Directive 4600.4, "Electronic Warfare (EW) and Command, Control, and Communications Countermeasures (C3CM)," Washington, D.C.: Department of Defense, August 27, 1979; Smith, 1983.

[42] Memorandum of Policy No. 30, "Command and Control Warfare," Washington, D.C.: Chairman of the Joint Chiefs of Staff, Joint Staff, July 17, 1990, revised March 8, 1993.

deception, electronic warfare, and physical destruction—into a single integrated C2W game plan. In a major change from previous doctrine, Gen. H. Norman Schwarzkopf added the strategy of attacking the entire Iraqi information system, including the human element through the fifth pillar of C2W—psychological operations. Because of its effectiveness during Desert Storm, command and control warfare . . . fostered fear and consternation among potential adversaries worldwide.[43]

U.S. strategy and tactics in the Gulf War played a significant role in shaping Soviet (soon-to-be-Russian) perceptions of U.S. capabilities and intentions.[44] This suggests that coalition efforts to influence, degrade, and destroy Iraqi C2 in the Gulf War also influenced Soviet and Russian military conceptions of modern warfare in the information age. Indeed, V. Slipchenko characterized the Gulf War as the "first example of information confrontation in a contactless war."[45]

Although Lyutov appears to have been the first in the Soviet military-scientific community to have published an article that used the term "information confrontation," he was not alone in acknowledging the growing role of information in combat, including in the targeting and protection of C2. The same year that Lyutov and Vayner penned their articles, their contemporary, General-Major I. N. Vorob'yev, made a similar observation—that confrontation between states was no longer exclusive to ground and air but was now also unfolding in radio and electronic military systems in what he described as a "struggle over the airwaves."[46]

---

[43] Hutcherson, September 1994, p. 4.

[44] Mary C. Fitzgerald, "The Soviet Image of Future War: Through the Prism of the Persian Gulf," *Comparative Strategy*, Vol. 10, No. 4, 1991, pp. 393–435.

[45] V. Slipchenko, "Novaya Forma Bor'by. V Nastupivsheme Veke Rol' Informatsii v Beskontaktnykh Voynakh Budet Lish' Vozrastat' [A New Form of Combat. In the Coming Century the Role of Information in Contactless Wars Will Only Increase]," *Armeiskii Sbornik [Army Digest]*, No. 12, 2002, p. 30.

[46] I. N. Vorob'yev, "Voprosy Teorii i Praktiki Manevrennoj Oborony [Questions of the Theory and Practice of a Mobile Defense]," *Voennaya Mysl' [Military Thought]*, No. 9, 1990, p. 36; Fitzgerald, 1991, p. 28.

Early Post–Cold War Era

Not long after these first articles were published, the international system underwent profound changes. The Soviet Union ceased to exist, and with it the Cold War was laid to rest. However, the thematic motifs expressed by Lyutov, Vayner, and Vorob'yev survived the dissolution of the Soviet Union and remained important fixtures in early Russian post–Cold War military-scientific thought. Like before, the community's publications continued to underscore the centrality of information (and its control) in modern combat and its role as a determinant of victory in the "post-nuclear era."[47] Likewise, much of the literature published in the wake of the Soviet collapse commented on the ever-increasing computerization and automation of the military sphere and the implications of this phenomenon for the Russian military in both wartime and peacetime.

In a 1992 article on intelligent C2 systems, Colonel-General A. P. Elkin and Colonel (Ret.) A. I. Starikov characterized IPb as a contest among states for information superiority. The state that was able to collect, process, and transmit information more quickly and establish control over vital information possessed a significant advantage. This, they argued, was particularly true in the dawning age of high-precision weapons, which significantly compressed decision time scales.[48] Like Vayner, Elkin and Starikov's threat perceptions were informed by U.S. and NATO behavior. Western states, they argued, had recognized the importance of information superiority in modern warfare and devoted resources toward developing what the authors termed "intelligent command and control systems" in response.[49] Elkin and Starikov cited the coalition partners' successful employment of high-precision systems—enabled by computer-facilitated

---

[47] A. I. Pozdnyakov, "Informacionnaya Bezopasnost' Lichnosti, Obshestva, Gosudarstva [Information Security of People, Society, and the State]," *Voennaya Mysl' [Military Thought]*, No. 10, 1993, p. 13.

[48] A. P. Elkin and A. I. Starinkov, "K Voprosu ob Intellektual'nykh Sistemakh Upravleniya [On the Question of Intelligent Control Systems]," *Voennaya Mysl' [Military Thought]*, No. 1, 1992, pp. 35–39.

[49] Elkin and Starinkov, 1992, p. 35.

rapid information collection, processing, and dissemination—in the Gulf War as evidence. Along similar lines, in a 1993 article, Colonel A. I. Pozdnyakov acknowledged that superior awareness was a "condition for victory" on the modern battlefield.[50]

The effect of the informatization of warfare became apparent during the First Chechen War, from 1994 to 1996, when senior military leaders noted that although psychological operations had been successful, psychological units seemed to lack experience and capabilities related to newer mediums of influencing target audiences, such as television.[51] In its aftermath, Y. Kozhuk noted that the conflict in Chechnya had taken place "in the context of a rapidly progressing process of informatization and computerization of the modern world."[52] Chechen separatists, he explained, had used information "to compensate for their insufficient military strength [by] organizing a broad information (and disinformation) offensive."[53] This information offensive sought to present the Chechen separatist movement as "national liberation movement" fighting an imperial power."[54] One commentator wrote that the Chechen conflict, in retrospect, had marked Russia's entrance "into a global information confrontation."[55]

---

[50] Pozdnyakov, 1993, p. 13.

[51] "Deistviia Soedninenii Chastei i Podrazdelenii SV Pri Provedenii Spetsial'noi Operatsii po Razoruzheniiu NVF v 1994–96 gg na Territorii Chechenskoi Respubliki Spisok Sokrashchenii i Abbreviatur [Actions of Divisions and Subdivisions of the Army During a Special Operation to Disarm Illegal Armed Groups in 1994–96 on the Territory of the Chechen Republic. List of Abbreviations and Acronyms]," *Doklad Byvshevo Nachal'nika Shtaba SKVO General-Leitenant V. Potapova [Report of the Former Chief of the North Caucasus Military District, Lieutenant-General V. Potapov]*, Vestnik PVO, April 11, 2005. As of June 7, 2021: http://pvo.guns.ru/book/chechnya_pvo.htm.

[52] Y. Kozhuk, "Zametki Voyennogo Obozrevatelya: Arena Informatsionnogo Protivoborstva [Notes of a Military Observer: Information Confrontation Arena]," *Flag Rodiny [Flag of the Motherland]*, No. 124, 2002a.

[53] Kozhuk, 2002a.

[54] Kozhuk, 2002a. It also included the production of "full-length feature films on Chechen anti-Russian themes."

[55] D. Makarov, "Informationnyye Voyny. Slovo, Postavlennoye pod Ruzh'ye [Information Wars. The Word, Placed at Gunpoint]," *Flag Rodiny [Flag of the Motherland]*, No. 115, July 4, 2009.

Throughout the 1990s, information operations that made use of evolving communication technologies particularly impressed Russian defense officials. As the former First Deputy Chairman to the State Technical Commission, Colonel-General Evgeniy Balyaev, claimed in a 1998 interview, Russia's intelligence adversaries continued to increase their spending and capabilities despite the end of the Cold War, with the United States allegedly spending $30 billion on these activities.[56] The journal *At the Fighting Post* (*Na Boevom Postu*) effectively summarized the situation in an article titled "War of the Future: The West Goes on an 'Information Offensive'":

> In the West, they reckon that thanks to the combination of modern information technologies, it will be possible to bring to a new level the concept of intimidation and deterring a likely enemy from attack. . . . This is well understood in the United States, whose attempts to achieve complete superiority in this specific area of military-technical progress are becoming increasingly obvious. It is the Pentagon that plays the main role in creating the theory of information warfare . . . in essence, preparations for the conduct of the information war in the West have already begun. And it is likely that in the coming years it will become much wider.[57]

NATO operations not only in the Persian Gulf but also in the Balkans and, later, in Afghanistan and Iraq did much to exacerbate Russian fears of a post-Soviet disparity in information operations. For example, during the bombing campaign against Serbian forces, NATO's use of "Commando Solo," a C-130 with a unique electronic warfare platform, captured many mid-echelon Russian officers' attention, which evidently proved to some NATO's desire to not only achieve "physical destruction," but to establish dominance over a state's "information

---

[56] "Natsional'naia Bezopasnost', Informatsiia—Tozhe Oruzhie [National Security: Information Is Also a Weapon]," *Vestnik Voennoi Informatsii [Journal of Military Information]*, No. 10, 1998.

[57] Elena Suvortseva, "Voiny Budushchego. Zapad Perekhodit v 'Informatsionnoe Nastuplenie' [Wars of the Future. The West Starts an 'Information Attack']," *Na Boevom Postu [At the Fighting Post]*, No. 42, 1997.

infrastructure" as well.[58] The Russian state simply lacked the resources to match these emerging U.S. and NATO capabilities.

A number of members of the Russian military-scientific community expressed major concern over the asymmetry they perceived to exist between Western and Russian capabilities in the ongoing information confrontation. Informational weapons, they warned, could pose a grave threat to Russia. The use of a decapitating electronic warfare strike on Russian C2 and communications systems, resembling the coalition operation in the Persian Gulf, was of particular concern. Remarks by the then-director of the 46th Central Research Institute (the GRU body responsible for assessing foreign states' potential), Colonel V. I. Tsymbal, speak to these anxieties.[59] In his speech at a 1995 conference, Tsymbal warned,

> [T]he use of information warfare means against Russia or its armed forces will categorically not be considered a non-military phase of a conflict, whether there were casualties or not . . . considering the possible catastrophic consequences of the use of strategic information warfare means by an enemy, whether on economic or state command and control systems, or on the combat potential of the armed forces, . . . Russia retains the right to use nuclear weapons first against the means and forces of information warfare, and then against the aggressor state itself.[60]

It bears noting that Tsymbal's use of the term "information warfare" (*informatsionnaya voyna*) in place of "information confrontation" (*informatsionnoe protivoborstvo*) was not uncommon in the early post–Cold War period. In the absence of a standardized lexicon for these concepts in the 1990s, the constructs put forward by military-scientific

58 Evgenii Grigorevich Zushin, "Vlast', ne Imeyushchaia Ravnykh po Sile Vozdeystviia [Power, Unparalleled in Terms of Impact]," *Nezavisimoe Voennoe Obozrenie [Independent Military Review]*, No. 16, 1999.

59 For information about the 46th Central Research Institute, see Vladimir Ivanov, "Samye Informirovannye Ludi v GRU [The Most Informed People in the GRU]," *Nezavisimaya Gazeta [Independent Gazette]*, October 12, 2012.

60 Timothy L. Thomas, "Russian Views on Information-based Warfare," *Airpower Journal*, July 1996.

theorists of this period used these as well as other adjacent terms, such as information struggle (*informacionnaia borba*).[61]

In a 1996 article, Colonel S. A. Komov shared his contemporaries' disquiet over the growing importance of IPb, or information struggle, in modern warfare. Though militaries have long employed informational instruments in combat, Komov acknowledged, these practices were typically limited to intelligence and counterintelligence and were usually of secondary importance, conducted in service of other operational ends. This changed, Komov argued, with the overwhelming informatization and automation of military hardware and software. For the first time, militaries could cripple an adversary by targeting their C2 systems and processes, before ever engaging in direct combat.[62]

Like his contemporaries, Komov cited Western behavior as evidence of these developments. The United States, he noted, had allocated significant resources to this issue, including establishing units devoted to the conduct of offensive information campaigns against adversary command and control.[63] He urged the Russian military-scientific community to keep pace with these developments. Specifically, Komov identified the need to better understand the "laws and principles of information warfare, [and] the active development of its scientific theory," as "an extremely urgent problem," the implication being that this issue was not yet a priority for the Russian military-scientific community.[64]

The same year, one of Russia's leading experts on information security issues, Major General E. G. Korotchenko, contributed to the

---

[61] Writing in 1996, specialist on Russian military issues Timothy Thomas noted that "while no official (that is, MOD endorsed) military definition of information was found in the research for this article, several unofficial ones were uncovered." Thomas writes that as of 1996, *informatsionnaya voyna* was the preferred lexicon among civilian circles whereas he described *informatsionnoe protivoborstvo* as "already in use by some military sources, to include the General Staff Academy," according to a source he spoke with. See Thomas, 1996.

[62] S. A. Komov, "Informacionnaya Bor'ba v Sovremennoj Vojne: Voprosy Teorii [Information Struggle in Modern Warfare: Theoretical Issues]," *Voennaya Mysl' [Military Thought]*, 1996.

[63] Komov, 1996, p. 76.

[64] Komov, 1996, p. 76.

burgeoning discourse on information confrontation.[65] Much of the relevant dialogue up to that point had focused on the more technical aspects of IPb, but Korotchenko underscored the importance of what he termed "informational-psychological confrontation," a relatively broad concept that amalgamated elements of the earlier constructs of psychological warfare and propaganda.[66] Existing psychological vulnerabilities, including rising rates of mental illness among the Russian population, served as fertile ground for adversary exploitation. Korotchenko predicted that adversaries like the United States would attempt to influence the perceptions and attitudes of Russian leadership, the public, and military personnel by leveraging foreign media and other soft power tools.[67] Without expressly stating so, he insinuated that the Russian mass media was being used as an instrument of influence, citing its dissemination of untruths as evidence.[68]

The assertive tone of Korotchenko's article represented a departure from previous discourse on the subject. Adversary "attempts to weaken and even destroy Russian statehood" through the employment of informational weapons, Korotchenko argued, were among the

---

[65] For additional background on Major-General Korotchenko, see "Pozdravliaem Yubilyarov [Happy Anniversary]," *Voennaya Mysl' [Military Thought]*, No. 12, December 2017, p. 83.

[66] E. G. Korotchenko, "Informacionno-Psihologicheskoe Protivoborstvo v Sovremennyh Usloviah [Informational-Psychological Confrontation in Modern Conditions]," *Voennaya Mysl' [Military Thought]*, 1996, pp. 25–26.

[67] Korotchenko, 1996, pp. 25–26.

[68] It is worth noting that this narrative mirrors motifs expressed in recent Kremlin rhetoric, which has in some cases attempted to discredit unflattering news by labeling it as fabricated information. A recent example of this phenomenon was a Kremlin effort to discredit reporting on COVID-19 related deaths in Russia as falsified. See "Fake News or the Truth? Russia Cracks Down on Virus Postings," *Associated Press*, April 1, 2020. Similarly, Russian authorities have recently undertaken a number of legislative efforts to restrain domestic media as well as other expressions of freedom. For instance, the regime has passed legislation that criminalizes any expression of "disrespect" toward Russian culture or the Russian authorities, or the dissemination of any content that are deemed to be fallacious. In a similar vein, the Russian government has also approved a law that allows the government to designate media organizations that are recipients of foreign funding as "foreign agents." See Alina Polyakova, "The Kremlin's Latest Crackdown on Independent Media," *Brookings*, December 6, 2017; Anna Nemtsova, "Putin's Crackdown on Dissent Is Working," *The Atlantic*, March 22, 2019; Andrew Higgins, "As Putin Era Begins to Wane, Russia Unleashes a Sweeping Crackdown," *The New York Times*, October 24, 2019.

leading threats facing Moscow.[69] He explicitly implicated the United States, which he accused of conducting information aggressions against Russia. Whereas contemporaneous literature principally called for the study of IPb, Korotchenko proposed that the growing informational threat be met with "necessary . . . retaliatory measures," a potential indicator of heightened Russian threat perceptions.[70]

In the waning years of the 20th century, information confrontation (and its many aliases) had "in a short period of time . . . become a priority" topic in the Russian military-scientific discourse.[71] Much of this literature classified IPb as either broad or narrow, with the former being a struggle among states for influence using informational instruments in peacetime and war, while the latter suggested a military contest for information superiority in advance of or during hostilities.[72] In any event, there was growing recognition that the "quality and volume of information potential" was becoming one of the "most important indicators of a state's defense capability," with military success "largely depend[ing] on victory in information confrontation."[73]

The Putin Era: 2000 to Present

The turn of the century marked another important juncture in the development of the concept of information confrontation. In the twilight of

---

[69] Korotchenko, 1996.

[70] Korotchenko, 1996, p. 23.

[71] M. A. Rodionov, "K Voprosu o Formakh Vedeniya Informatsionnoy Bor'by [On the Question of the Ways of Waging Information Warfare]," *Voennaya Mysl' [Military Thought]*, No. 2, 1998, p. 67; V. S. Pirumov and M. A. Rodionov, "Nekotorye Aspekty Informacionnoj Bor'by v Voennyh Konfliktah [Some Aspects of Information Struggle in Military Conflicts]," *Voennaya Mysl' [Military Thought]*, No. 5, 1997, p. 45.

[72] As noted above, there were theoretical differences across the writings of different authors in this era. In their 1997 piece, Pirumov and Rodionov, make the distinction between broad and narrow informational efforts, but assign the term "information confrontation" to mean high-level geopolitical competition between states (broad) and the term "information warfare" to mean efforts specific to the military. See Pirumov and Rodionov, 1997, p. 45; Komov, 1996.

[73] A. Markov, "Informatsionnoye Obespecheniye [Information Support]," *Armeiskii Sbornik [Army Digest]*, No. 11, November 1997.

the Boris Yeltsin period, the administration published several new strategic-level documents, which marked the first time that IPb was incorporated into Russian national security strategy. The Kremlin's January 2000 *National Security Concept* warned of the "increased threat to the national security of the Russian Federation in the information sphere."[74] The document cited "the desire of a number of countries to dominate the global information space and to expel Russia from the external and internal information market," and emphasized that the "'information warfare' concepts that are developed by these states which entail the creation of means to exert a dangerous impact on other countries' information systems" pose serious threats to Russia.[75] The tone of the document represented a departure from the 1997 *National Security Concept*, which had acknowledged the increasing significance of information security as an element of Russian national security but focused on the necessity of making infrastructure and technical improvements in the information sphere.[76]

The 2000 Russian *National Security Doctrine* struck a similar tone. The document explicitly named the "exacerbation of information confrontation" as one of the main factors influencing Russia's military-political situation (VPO).[77] States' use of "informational and other (including non-traditional) means and technologies for aggressive (expansionist) purposes," it noted, contributed to the destabilization of the military-political situation.[78] The discussion of information confrontation in leading strategy documents at the turn of the century indicates that the concept had, in somewhat short order, become a priority issue among upper echelons of Russian military and political elites.

---

[74] Russian Federation, *Russian National Security Concept*, January 10, 2000.

[75] Russian Federation, 2000.

[76] See Table 1.1 for a definition of "information sphere," as defined in the Russian literature. Russian Federation, "Russian National Security Blueprint," *Rossiiskaya Gazeta [Russian Gazette]*, December 26, 1997, pp. 4–5.

[77] Decree of the President of the Russian Federation, "On the Approval of the Military Doctrine of the Russian Federation," N 706, April 21, 2000.

[78] Decree of the President of the Russian Federation, 2000.

As might be expected, scholarship devoted to information confrontation proliferated in the years following its official designation as a national security priority. In the early 2000s, the Russian military-scientific community began to explore the nuances of IPb, including how it unfolded in specific military domains, how it affected other means and methods of warfare, and how it could be used to influence specific audiences. In 2002, for example, reserve Colonel G. S. Chernykh examined the effects of IPb on radiological, chemical, and biological (RCB) missions, concluding that information security was critical to the operations of these troops given the chaos that would result from the dissemination of fabricated messages about RCB weapons.[79] He went on to identify vulnerabilities in the Russian systems that RCB troops used to communicate; if exploited, these systems could be used to distort the military and media's understanding of the situation on the ground.

Other scholars analyzed the specific effects of U.S. information confrontation activities on Russian troops and command and control systems. In 2003, for example, Major General (Ret.) A. N. Limno and reserve Colonel M. F. Krysanov advocated for the reevaluation and reengineering of the Russian *maskirovka* (referred to as military deception in Western circles) practices in order to conceal and harden Russian C2 against foreign influence, as well as to establish more effective means to deceive the adversary.[80]

As the decade progressed, new geopolitical, economic, and technical developments, including the maturation of the internet and NATO operations in Kosovo, influenced evolving Russian conceptions of information confrontation.[81] In a 2008 article, D. P. Prudnikov

---

[79] G. S. Chernykh, "Ob Uchastii Vojsk RHB Zashity v Informacionnom Protivoborstve [On the Participation of RCB Protection Troops in Information Confrontation]," *Voennaya Mysl'* [*Military Thought*], No. 6, 2002, pp. 47–49.

[80] Limno and Krysanov, 2003, pp. 70–74.

[81] For instance, R. Bikkenin acknowledges the centrality of the internet in modern information confrontation. See R. Bikkenin, "Informatsionnoye Protivoborstvo v Voyennoy Sfere [Information Conflict in the Military Sphere: Basic Elements and Concepts]," *Morskoi Sbornik* [*Naval Digest*], No. 10, 2003.

noted that the history of warfare in the late 20th and early 21st centuries "testifies to the unconditional growth of the role of the information factor" in armed conflict.[82] Information, Prudnikov stated, had become the "most important military-strategic resource, no less or even more important than traditional types of weapons."[83]

Although Russian military scholars had characterized the Gulf War as the first true information war, Kosovo, according to Colonel Yu. O. Yashchenko, was the "first manifestation of information warfare on the Internet."[84] According to Yashchenko, the Kosovo War served as a bellwether for internet-based information aggressions. These activities spanned the gamut from highly technical denial-of-service attacks designed to disable adversary email to psychological operations conducted online to malign adversaries in the eyes of regional and global audiences. For instance, Yashchenko cited what he claimed was NATO's use of "American news agency" CNN to create online content designed to vilify the Serbs as perpetrators of "ethnic cleansing" and "massacres," implying this language was an inaccurate representation of events.[85] This content, Yashchenko suggested, had been generated by the 30th battalion of the U.S. Army's 4th Psychological Operations Group, working in concert with CNN.[86]

Yashchenko's observations of NATO operations in Kosovo informed his development of a schema for categorizing information confrontation activities conducted using the internet. He identified four types of IPb: (1) the deliberate dissemination of disinformation

---

[82] D. P. Prudnikov, "Gosudarstvennaya Informatsionnaya Politika v Oblasti Oborony: Iskhodnoye Opredeleniye [State Information Policies in the Defense Sphere: Initial Definition]," *Voennaya Mysl' [Military Thought]*, Vol. 3, 2008, p. 47.

[83] Prudnikov, 2008, p. 43.

[84] Yashchenko, 2003, p. 72.

[85] Yashchenko, 2003, pp. 72–73.

[86] Yashchenko, 2003, p. 76. Other theorists in the military-scientific community expressed similar views about Western media organizations operating as instruments of military and foreign policy. See, for instance, S. G. Chekinov and S. A. Bogdanov, "Strategicheskoe Sderzhivanie i Nacional'naa Bezopasnost' Rossii na Sovremennom Etape [Strategic Deterrence and National Security, Russia in the Modern Period]," *Voennaya Mysl' [Military Thought]*, No. 3, 2012, p. 18.

by email and established online opinion forums, and by posting online versions of traditional news; (2) hacking websites and replacing legitimate content with fabricated or new content; (3) offensive cyberattacks designed to harm internet functionality; and (4) email bombing with the intention of disrupting legitimate email services.[87] This construct captured activities of both a technical and psychological nature. The same year, several of Yashchenko's contemporaries discussed analogous concepts, which they referred to as the informational-technical and informational-psychological facets of IPb.[88]

Yashchenko was not alone in viewing Kosovo as instructive for the future of information warfare. In 1999, one commentator remarked that pro-Milosevic forces were defeating NATO through information operations "despite the situation in the air and on the ground." Milosevic's supporters used the internet and satellite broadcasting to influence European target audiences regarding the ongoing conflict. Further, NATO bombing of Serbian relay towers nullified Western television broadcasts in the region and increased the prominence of internet-based operations.[89]

In addition to Kosovo, U.S. and NATO military campaigns in Afghanistan and Iraq in the early years of the 21st century provided Russian military observers and scholars with many lessons about information warfare. Russian military experts perceived that during the U.S. and NATO invasion of Afghanistan, the Taliban had outperformed NATO in terms of information warfare because of the latter's cultural unfamiliarity and difficulty working with local partners, while the Taliban deftly used religion, a wide intelligence network, and disinformation on the

---

[87] Yashchenko, 2003, pp. 72–78.

[88] See S. P. Rastorguyev, "An Introduction to the Formal Theory of Information Warfare," Moscow, 2003, cited in Timothy L. Thomas, *Comparing US, Russian, and Chinese Information Operations Concepts*, Fort Leavenworth, Kan.: Foreign Military Studies Office, February 2004, pp. 6–9; S. A. Bogdanov, "Veroyatnyj Oblik Vooruzhennoj Bor'by Budushego [Likely Features of the Armed Struggle of the Future]," *Voennaya Mysl' [Military Thought]*, No. 12, 2003, p. 5; R. Bikkenin, 2003.

[89] "Na Informatsionnom Fronte, Est' li u Miloshevicha Svoi Udugov? [On the Information Front, Does Milosevic Have Own Udugov?]," *Soldat Otechestva [Soldier of the Motherland]*, No. 39, 1999.

activities of foreign forces.[90] Writing in 2003, then–Vice Admiral E. I. Volobuyev observed that "information confrontation [had] become an integral part of influencing all types of military combat actions, including fires." He had studied coalition operations in Iraq and the former Yugoslavia closely and concluded that his colleagues in the Russian military-scientific community had not gleaned the right lessons from these conflicts.[91] He was not alone in expressing this sentiment. Writing in 2008, security services veteran-turned-scholar I. Panarin lamented,

> Information wars are a real factor in geopolitics. By this term, we usually mean a whole range of actions to influence the psyche and behavior of individuals and society as a whole, as well as measures to combat this influence. Unfortunately, this factor is underestimated by the political elite of Russia, as before it by the USSR. And, this underestimation is precisely the root of many issues facing modern Russia.[92]

Like many of his contemporaries, Panarin drew lessons from the 2008 Russo–Georgian War, the latest conflict to unfold at the time of his writing. From Panarin's vantage point, the conflict served as a glaring example of Russia's failure to conduct a coordinated information campaign to support Russian interests in global information space— the kind of effort the West had successfully executed.[93] In the wake of the Russo–Georgian War, Russian political elites convened expert panels to discuss lessons learned from the conflict, according to Panarin. The specialists agreed that above all, Russian authorities had seriously underestimated the role of information confrontation in modern interstate competition.[94]

---

[90] A. Novik, "Stavka na Spetsoperatsii [Bet on Special Operations]," *Strazh Baltiki [Guardian of the Baltic]*, No. 104, 2009.

[91] E. I. Volobuyev, "VMF i Problemy Kompleksnogo Ognevogo Porazhenia Protivnika [Navy and Problems of Defeating the Adversary through Complex Fires]," *Voennaya Mysl' [Military Thought]*, No. 4, 2003, pp. 26–27.

[92] I. Panarin, "Sistema Informacionnogo Protivoborstva [System of Information Confrontation]," *Voenno-Promyshlennii Kur'er [Military-Industrial Courier]*, No. 14, October 14, 2008.

[93] Panarin, 2008.

[94] Panarin, 2008.

According to Panarin, Russia's future success in the information age would hinge on its ability to devise and execute a coherent approach to information confrontation. While his approach largely followed in the vein of previous scholarship on the subject, it diverged from existing work in its proposals for specific policy changes as well as its recommendation to establish a system of structures responsible for carrying out IPb. A number of the changes Panarin recommended—notably, increasing the centralization of internal and external communications of all state organs and private business and subsuming media outlets like RT under the Russian Foreign Ministry—paralleled developments that have transpired in the years since the publication of his article.[95] In 2013, moreover, the new defense minister of Russia, Sergey Shoygu, announced a "big hunt" to incorporate highly qualified researchers and technical specialists into the ranks of the military, which received presidential approval and had a fairly transparent emphasis on cyber capabilities.[96] According to Shoygu, this force will aim to ensure that Russia is more effective in counter-propaganda.[97] According to open-source reporting, the main tasks of the new force are coordinating and integrating operations carried out in cyberspace, as well as the management and protection of military computer networks.[98]

The Russian military's 2015 intervention in the Syrian conflict further solidified the role of information confrontation in contemporary

---

[95] Panarin, 2008.

[96] Probably not coincidentally, a Russian state-sponsored fictional TV program accompanied the recruiting initiative that featured young hackers and computer science specialists serving in a new military unit. Daniil Turovskii, "Rossiyskie Vooruzhenye Kibersily kak Gosudarstvo Sozdaet Voennye Otryady Khakerov [Russian Armed Cyber Forces: How the State Creates Military Hacker Units]," *Meduza [Medusa]*, November 7, 2016.

[97] Ivan Petrov, "Shoigu Announced the Creation of Information Operations Troops [Shoigu Obyavil O Sozdanii Voisk Informacionnih Operacii]," *Rossiskaya Gazeta*, February 22, 2017.

[98] General Leonid Ivanshov, president of the Academy of Geopolitical Problems, and other Russian academics have emphasized the need for Russia to counter what they perceive as Western lies and offensive information operations. General Ivanshov has also suggested that Russia needs to create a state-level headquarters that engages not only in counter-propaganda, but also carries out informational-psychological offensive operations and "convey[s] our truth to the leaderships of the countries of the world." Peter Menchikov, "Information Battlefield [Informacionnoye Pole Boya]," *Nacionalnaya Oborona*, November 2020.

conflict. The commander of Russian ground forces during the Syria campaign drew the following conclusions from operations during the fighting around Aleppo, which centered on information operations and nonlethal effects:

> In the course of operations in Syria we, like nowhere before, were convinced of the practicality of information confrontation. Information resources have become in fact one of the most effective types of weapons. Their widespread use allows you to stir up the . . . situation from the inside in a matter of days. For example, during the operation to liberate Aleppo, informational work with the local population helped to liberate entire neighborhoods without a fight, to bring out more than 130 thousand civilians.[99]

---

[99] Aleksandr Dvornikov, "Shtaby Dlya Novykh Voyn [Headquarters for the New Wars]," *Voenno-Promyshlennii Kur'er [Military-Industrial Courier]*, No. 28, 2018. In 2015, even before Russian military intervention in the conflict, the GRU launched an extended information campaign through a proxy dubbed "CyberCaliphate," probably at least in part to galvanize Western governments and publics against the threat posed by the Islamic State in Syria. See also "Reckless Campaign of Cyber Attacks by Russian Military Intelligence Service Exposed," *National Cyber Security Centre*, October 3, 2018.

CHAPTER THREE

Information Confrontation in Russian Strategy

The concept of information confrontation should be considered within the broader context of Russian national security and defense strategy, which describes Russia's domestic and global ambitions, threat perceptions, and national identity, as well as the Kremlin's view of Russia's place in the world. Russia's understanding of its role in the global information space is closely linked to its understanding of its place in the world. Russia views information confrontation not only as a threat—in both the domestic and global information space—but also as a means of achieving its political and strategic objectives.[1]

This chapter provides background on Russian national security and defense strategy before turning to an examination of key Russian strategic documents addressing the role of information and information technologies in national security, strategic competition, and active conflict. It then considers how the central role of IPb in Russian strategy has shaped Russian views on the use of international law in the regulation of information weapons and information operations.

---

[1] Russian strategic documents and authors apply a broad and inclusive understanding of information space (which includes both information technology as the means and information as the contents), and confrontation in the information space. See Andrei Soldatov and Irina Borogan, "Russia's Approach to Cyber: The Best Defence Is a Good Offense," in *Hacks, Leaks and Disruptions: Russian Cyber Strategies*, European Union Institute for Security Studies, October 1, 2018, pp. 15–24; Emilio J. Iasiello, "Russia's Improved Information Operations: From Georgia to Crimea," *Parameters*, Vol. 2, No. 47, 2017, p. 61.

Overarching Themes

Russia's Foreign Policy Concept (2016) asserts that the United States and its allies have adopted a "containment policy" and are exerting pressure against Russia and other countries in many spheres, including the information sphere. This effort is aimed at expanding the geographic reach of Western influence and maintaining the United States' dominant position in the world:

> The containment policy adopted by the United States and its allies against Russia, and political, economic, information, and other pressure Russia is facing [as a result] undermine regional and global stability, [and these factors] are detrimental to the long-term interests of all sides and run counter to the growing need for cooperation and addressing transnational challenges and threats in today's world.[2]

Russian perceptions are also influenced by the belief that the United States and its allies seek to dominate the information sphere through both cognitive and technical means and thereby eliminate, or at least reduce, the international influence of Russia in the global information space and reduce Russia's ability to control its domestic information sphere. The development of information warfare concepts and capabilities by the United States and its allies is used as evidence in support of the assertion that the West seeks to undermine the governance systems of its adversaries and act as a destabilizing influence on international relations.[3] Russian Defense Minister Shoygu, for example, has accused the West of waging information wars against Russia with the

---

[2]   Ministry of Foreign Affairs of the Russian Federation, *Foreign Policy Concept of the Russian Federation*, approved by the President of the Russian Federation on November 30, 2016, released December 1, 2016.

[3]   Russian Federation, *Doktrina Informatsionnoy Bezopasnosti Rossiyskoy Federatsii [Information Security Doctrine of the Russian Federation]*, December 5, 2016.

aim of preventing Russia from assuming its rightful position as one of the geopolitical poles of the world.[4]

Russia sees itself as engaged in an ongoing information confrontation with the West, and specifically with the United States.[5] Historically, Russia—and its predecessor, the Soviet Union—has been viewed as being the eternal, irreconcilable rival of the United States.[6] An article published by the Russian Ministry of Internal Affairs claims that the concept of "rivalry," or *sopernichestvo*, was formulated by the United States. Within the context of this rivalry, the struggle to maintain and disseminate one's own values and cultural norms is a key element of the informational-psychological aspect of IPb. As A. A. Irkhin explains, "At the heart of the modern competition of the leading powers of the world are two main areas of struggle—the struggle for resources and

---

[4] A. M. Tsygankov complements this by arguing that foreign actors specifically aim to gain control over the consciousness of Russia's youth. Ministry of Defense of the Russian Federation, "Ministr Oborony Sergey Shoygu Nazval Glavnoy Tsel'yu Informatsionnoy Voyny Zapada Protiv Rossii Polnoye Yemu Podchineniye [Defense Minister Sergey Shoygu Called Complete Submission to the West as the Main Goal of the Information War of the West Against Russia]," June 26, 2019; A. M. Tsygankov, "Voyenno-Politicheskiye Aspekty Stroitel'stva i Razvitiya Vooruzhennykh Sil Rossiyskoy Federatsii na Sovremennom Etape [Military and Political Aspects of Construction and the Development of the Armed Forces of the Russian Federation on the Modern Stage]," *Voyenno-Nauchnaya Konferentsiya Akademii Voyennykh Nauk [Military-Scientific Conference of the Academy of Military Sciences]*, Vol. 67, No. 2, 2019, p. 20.

[5] A. A. Efanov, "Izvechnoye Rossiysko-Amerikanskoye Protivostoyaniye, ili O Chetyrekh Etapakh Odnoy Informatsionnoy Voyny v Noveyshey Istorii [The Eternal Russian-American Confrontation, or About Four Stages of One Information War in Modern History]," in *Informatsionniye Voyny kak Bor'ba Geopoliticheskikh Protivnikov, Tsivilizatsii i Razlichnykh Etosov [Information Wars as Struggle Between Geopolitical Opponents Civilizations and Ethos: Collection of Works of All-Russian Scientific Conference]*, Novosibirsk: Siberian State University Telecommunications and Information, April 26–27, 2018, pp. 217–224.

[6] A. A. Efanov traces this rivalry to the slogan "catch-up and overtake" and Lenin, who wrote, "Either perish, or catch up with the advanced countries and overtake them also economically. Efanov, 2018, p. 217. See also V. I. Lenin, *The Impending Disaster and How to Deal with It [Grozyashaya Katastrofa I Kaka S nei Borotsa]*, 1917. V. V. Ovchinnikov and N. M. Petrovich, "Information Confrontation in Modern Geopolitics [Informacionnoye Protivoborstvo V Covrememmoi Geopolitike]," Russian Ministry of Internal Affairs Joint Editing Office, undated.

competition for meaning and ideas that exert power over the minds of the people and entire civilizations."[7] According to Sergey Pyastolov, "Information wars are never cold, but in these wars, people are programmed, not killed."[8] Likewise, A. A. Efanov compared the nature of information wars and total wars, explaining that the ultimate goal of an information war is to replace, displace, or integrate sociocultural values. This process does not act as a total replacement for active conflict, as exerting sociocultural influence over an adversary may be used both as a means of conflict prevention as well as a way to prepare the operational environment for a future military conflict.[9]

While Russian military scholars assert that Russia—and previously, the Soviet Union—has been in an information confrontation for decades, in recent years, they have emphasized that Russia and the United States are in an active phase of IPb. According to A. A. Efanov, for example, Russia has been in an active phase of IPb with the United States since the late 2010s. He predicts that this confrontation will continue to aggravate the global geopolitical situation and threaten a potentially dangerous transition to military engagement.[10]

Information Confrontation in Peacetime and Competition

According to Russian Military Doctrine, the world—both at present and in the future—will be characterized by a "struggle for dominance

---

[7]   A. V. Stavickii, "Ontologicheskiye Osnovy Informatsionnoy Voyny v Kontekste Bolshoy Igry Protiv Rossii [Ontological Foundations of Information Warfare in the Context of the Great Game Against Russia]," in *Informatsionniye Voyny kak Bor'ba Geopoliticheskikh Protivnikov, Tsivilizatsii i Razlichnykh Etosov [Information Wars as Struggle Between Geopolitical Opponents Civilizations and Ethos: Collection of Works of All-Russian Scientific Conference]*, Novosibirsk: Siberian State University Telecommunications and Information, April 26–27, 2018, pp. 664–677.

[8]   Pyastolov, 2019, pp. 17–21.

[9]   V. M. Zolotuhin, "Sokhraneniye Sotsiokul'turnoy Rossiyskoy Identichnosti v Prostranstve Informatsionnykh Voyn [Preservation of Russian Sociocultural Identity in the Space of Information Wars]," in *Informatsionniye Voyny kak Bor'ba Geopoliticheskikh Protivnikov, Tsivilizatsii i Razlichnykh Etosov [Information Wars as Struggle Between Geopolitical Opponents Civilizations and Ethos: Collection of Works of All-Russian Scientific Conference]*, Novosibirsk: Siberian State University Telecommunications and Information, April 26–27, 2018, pp. 231–239.

[10] Efanov, 2018, p. 223.

in shaping the key principles of the future international system."[11] In this struggle, the Kremlin seeks to position Russia as a key power in a multipolar world and offer an alternative vision of the international system, values, and policies to that represented by the United States and its allies. As such, Russia seeks to establish itself as a protector of the world against the declining international legal order and the usurpation of the West. This framing means that IPb, which at its essence consists of countering actors in the information sphere who seek to violate the sovereignty of independent states, is an important element of state policy.[12]

For Russia, information confrontation is closely linked with its aims of preserving and asserting its sociocultural identity and protecting this identity from what it perceives to be overwhelming and potentially destructive foreign influences. In March 2021, Putin highlighted the strategic importance of strengthening Russian civic identity, noting that "nothing [was] more important for our country."[13] Russian strategists have characterized influencing national identity, values, and way of life as among the primary goals of IPb. According to V. M. Zolotuhin, IPb "aims to create coalitions and organize media campaigns in order to discredit the identity of a potential adversary by imposing their own way of life, cultivating their own values, [and] creating an 'image of the enemy' that resembles one's opponents."[14] As such, Russia seeks to promote the Russian language and culture and enable Russian media outlets to occupy a prominent position in the global information space, thereby counterbalancing the international information space, which is dominated by the English language, and ensuring that Russia has a medium through which it can disseminate its "perspective on international processes to the wider international community."[15] These objectives are related to the Kremlin's goal of ensuring the survival of

---

[11] Ministry of Foreign Affairs of the Russian Federation, 2016.

[12] Streltzov, 2013, p. 21.

[13] "Putin Nazval Klycheboi Temy Ukrepleniya Rossiskoi Grazdanskoi Identichnosti [Putin Called Strengthening Russian Civic Identity to Be Key Theme]," *RBC*, March 30, 2021.

[14] Zolotuhin, 2018, p. 233.

[15] Ministry of Foreign Affairs of the Russian Federation, 2016.

the current Russian regime and diminishing the appeal of Western-style democracy by promoting an alternate worldview and destabilizing the United States and its alliances.[16]

According to Russian strategists, information confrontation also serves as a means of weakening the adversary in advance of a military conflict. According to Chief of the General Staff General Valeriy Gerasimov, one of the "characteristic features of modern military conflicts is the destabilization of the internal security of the state by sabotage and terrorism," which necessitates the "elaboration of a territorial defense system [and] constant readiness."[17] Gerasimov echoes the view of early-20th-century Russian military scholar Svechin, who wrote that a single military strike is likely to be effective in defeating an adversary that has been weakened by an internal conflict, while otherwise victory can only be achieved through a war of attrition.[18] IPb can facilitate this process by weakening an adversary and bringing about favorable changes in an adversary's internal political processes.[19]

Information Confrontation in Military Conflict

Information confrontation is not only a means of engaging in strategic competition, but also a critical element of military conflict. K. A. Sayfetdinov distinguishes between the use of IPb in peacetime, competition, and wartime. In peacetime, he writes, the purpose of IPb is to achieve the objectives set by the country's political leadership. It serves to enhance the country's political, diplomatic, economic, and legal capabilities. During competition, IPb can be used to carry out tasks for both the military and political leadership, in support of military and nonmilitary activities and aims. Finally, during wartime, IPb should be used to "gain and maintain information superiority over the

---

[16] Margaret L. Taylor, "Combatting Disinformation and Foreign Interference in Democracies: Lessons from Europe," *Brookings*, July 31, 2019.

[17] Valery Gerasimov, "Vectory Razvitiya Voyennoy Strategii [Military Strategy Development Vectors]," *Krasnaya Zvezda [Red Star]*, March 4, 2019.

[18] Andrew Monaghan, "How Moscow Understands War and Military Strategy," *CNA*, November 2020.

[19] Streltsov, 2013, pp. 20–21.

enemy in order to create favorable conditions for the successful actions of groupings of troops (forces) as intended" by the country's military-political leadership.[20] Some scholars have suggested that IPb allows Russia to "win small wars and resolve military conflicts without using conventional weapons."[21] The former chief of the General Staff, General Yuri Baluyevsky, has also characterized IPb as vital for military victory, stating: "Victory over the enemy in this war [IPb] can be much more important than victory in a classic military confrontation, since it is bloodless, and the effect is striking, draining and paralyzing [for] all the authorities of the enemy state."[22]

Although Western commentators have stated that Russian military and strategic thinkers prefer measures short of war, these assessments, as Russia scholar Andrew Monaghan concludes, do not accurately represent Russia's understanding of war and military strategy.[23] The military doctrine notes the importance of the information space and information activities in achieving defense and military goals, but in general, Russian military scholars and officers view IPb as just one facet of Russia's geopolitical confrontation with its adversaries, albeit an integral one.[24] For example, during his 2019 annual speech at the Russian Academy of Military Sciences, Gerasimov stated that although modern wars include both military and nonmilitary measures, the armed forces still play the decisive role.[25] Despite this, the Russian military-scientific literature emphasizes that information is the primary nonmilitary means of waging war. M. A. Gareev and N. I. Turko, for example, note that informational means are the primary nonmilitary means during the

---

[20] Sayfetdinov, 2014, p. 39.

[21] S. Grinyaev, "Tochka Zreniya: Informatsionnoye Prevoskhodstvo Vmesto 'Yadernoy Dubinki' [Opinion: Information Superiority Versus the 'Nuclear Stick']," *Armeiskii Sbornik [Army Digest]*, No. 5, 2002, p. 77.

[22] "Information Troops Created in Russia [V Rossii Sozdani Voiska Infromacionnih Operacii]," *RIA Novosti*, February 22, 2017.

[23] Monaghan, 2020.

[24] Efanov, 2018; Tsygankov, 2019, p. 20.

[25] Gerasimov, 2019.

first phase of warfare.[26] Other military scholars have suggested that non-military means play a more prominent role in modern warfare, such that the "effect of information influence can be [considered] comparable to the results of large-scale employment of troops and forces."[27]

Information is an element of the "whole spectrum of instruments" that Russia uses to achieve certain military and nonmilitary effects.[28] According to the Military Doctrine of the Russian Federation, discussed in more depth in the next section, military engagement should be considered only after nonmilitary means, including political, diplomatic, legal, economic, informational, and other nonviolent means, have been exhausted.

Russian military scholars have written at length on the role and aims of information confrontation in the military sphere. According to K. A. Trotsenko, information can be used to achieve superiority in command and control, preempt enemy actions, and improve the effectiveness of reconnaissance, electronic warfare, tactical camouflage, fires, and highly mobile units. According to Trotsenko and N. A. Molchanov, the primary aim of IPb in the military sphere is to "reduce the effectiveness of command and control of enemy troops" while ensuring Russian superiority in command and control, protecting Russian

---

[26] M. A. Gareev and N. I. Turko, "Voyna: Sovremennoye Tolkovaniye Teorii i Realii Praktiki [War: Modern Interpretation of Theory and Realities of Practice]," *Vestnik Akademii Voyennykh Nauk [Journal of the Academy of Military Sciences]*, Vol. 1, No. 58, 2017, p. 5. According to E. A. Derbin, only the second phase of warfare—which consists of combat and special operations—is purely military. The phases are Preparation, Stage 1: Misleading the international community about one's aims; Stage 2: Combat and special operations activities and political demands; Stage 3: Change of political leadership. E. A. Derbin, "Metodicheskye Aspekti Sushnosti Sovremennykh Voin [Methodological Aspects of the Analysis of the Essence of Modern Wars]," *Vestnik Akademii Voyennykh Nauk [Journal of the Academy of Military Sciences]*, Vol. 1, No. 58, 2017, pp. 11–18.

[27] A. S. Fadeev, "Voyenniye Konflicty Sovremennosti, Perspektivy Razvitiya Sposobov ikh Vedeniya, Priamiye i Nepriamiye Deistviya v Kofliktakh XXI Veka [Military Conflicts of Modernity, Prospects for the Development of the Ways of Their Management. Direct and Indirect Actions in Armed Conflicts of the 21st Century]," *Voennaya Mysl' [Military Thought]*, November 7, 2019.

[28] Quentin E. Hodgson, Logan Ma, Krystyna Marcinek, and Karen Schwindt, *Fighting Shadows in the Dark: Understanding and Countering Coercion in Cyberspace*, Santa Monica, Calif.: RAND Corporation, RR-2961-OSD, 2019.

command and control systems, and achieving information superiority.[29] S. A. Modestov notes that an additional aim of IPb is to preserve the state's ability to prevent its adversaries from achieving information superiority. He suggests that this goal can be achieved in two ways: by directly influencing information processes and by exerting influence at the critical points of the process of armed conflict. This dual approach has both tactical and operational benefits, allowing for the "pre-empt[ion of the] enemy's actions."[30] Several Russian authors acknowledge that Russia has not always been successful in information operations, highlighting Russia's inability to formulate and carry out the informational component of military operations in the Chechen Republic and South Ossetia.[31]

K. A. Sayfetdinov characterizes the aims of information confrontation in the military sphere as twofold: first, it facilitates the achievement of information superiority over the armed forces of one's adversary, and second, it creates favorable conditions for the preparation and employment of military forces. Activities in the information sphere may serve to deter or prevent military conflicts, as well as to prepare and shape the operational environment for a future conflict. Sayfetdinov identifies the main tasks of IPb in the military domain as follows:

- monitoring information sources and the identification, assessment, and forecasting of threats;
- deceiving the adversary about one's plans and intentions;
- sowing disarray among military forces and within the command and control of the adversary's forces;

---

[29] Trotsenko, 2016, pp. 20–25; Molchanov, 2008, pp. 2–9.

[30] Modestov, 2009, pp. 1–2.

[31] See, for example, Sayfetdinov, 2014. A. A. Shevcov also claims that the lack of an information component in Russian military operations led to the situation on the ground being interpreted mainly through foreign media sources, which he claims presented distorted or even false information. This resulted in the global community reacting negatively to Russian actions. A. A. Shevcov, "Information Strategy on the Russian Federation Based on the Example of the Military Conflict in Syria," *Communicology: Electronic Scientific Magazine,* Vol. 3, No. 1, 2018, pp. 59–67.

- reducing the psychological stability of the enemy's population and military personnel;
- maintaining the stable moral and psychological state of one's own military personnel; and
- protecting automated command and control systems.[32]

At least some Russian authors view war as a temporary state in modern conflicts, where a constant state of confrontation may be interspersed with periods of nonmilitary pressure, unconventional means of confrontation, information confrontation, and military conflict. According to this view, military conflict does not have precisely defined start and end points.[33] In fact, according to one Russian military expert, modern wars consist of 80 to 90 percent propaganda and 10 to 20 percent violence; they "are never declared and never ended."[34] In the 21st century, war is essentially "unlimited in time, space, and number of participants; it is conducted in all spheres of society, at all levels (strategic, operational, tactical), and by all possible means."[35] In modern conflicts, the goal is to gain control over the adversary's leadership and elite, rather than seek to inflict physical damage or the destruction of infrastructure.[36] Informational means are an integral part of the progression from competition to warfare to the achievement of military and political aims. This progression begins with the aggressor applying nonmilitary pressure to its adversary and "increasingly applying an unconventional model of confrontation," which may include both direct and indirect actions, while also establishing the legitimacy of an impending military conflict in the eyes of the international community. This may be achieved, for example, by creating

---

[32] Sayfetdinov, 2014, p. 40.

[33] Fadeev, 2019.

[34] Fadeev, 2019; Y. A. Chizhevskii, "Osnovniye Tendentsii Transformatsii Prirody i Kharaktera Sovremennykh Voyenno-Politicheskikh Konfliktov [Main Trends in the Transformation of the Nature and Character of Contemporary Military-Political Conflicts]," *Voennaya Mysl' [Military Thought]*, July 11, 2020.

[35] Fadeev, 2019; Chizhevskii, 2020.

[36] Fadeev, 2019.

"seemingly irrefutable evidence of a threat to regional security" posed by the adversary, accompanied by an active information campaign that is aimed at swaying the international community.[37] A. S. Fadeev suggests, moreover, that information confrontation may have a more significant impact than the large-scale employment of forces, as evidenced by the Ukrainian national revival, Western support for the opposition in Syria, and the Arab Spring.[38]

The Russian military-scientific literature characterizes information confrontation as an element of asymmetric warfare, which is understood as providing a "strategic balance between the military superpowers."[39] As part of this asymmetric response, IPb seeks to identify and target the most vulnerable aspects of the adversary's capabilities, weapons systems, and infrastructure. As a result, IPb is an economical means of enabling Russia to counter the superior combat capabilities of the United States.[40] Information confrontation helps to undermine the foundations of the state and can even effect regime change in the adversary's country. The most effective ways of conducting IPb, according to one Russian military strategist, are "falsification, the substitution of information, or its distortion."[41] Information confrontation reduces the military resources that are necessary to achieve a specific foreign policy effect and serves as a cost-efficient means of preparing the environment for potential military action. It also creates plausible deniability. As one Russian military scholar explains, the "use of indirect actions and methods," such as informational means, "makes it possible to achieve

---

[37] Fadeev, 2019.

[38] Fadeev, 2019.

[39] V. V. Selivanov and Y. D. Ilin, "Metodika Kompleksnoy Podgotovki Asimmetrichnykh Otvetov pri Programmno-Tselovom Planirovanii Razvitiya Vooruzheniya [Methodological Foundations for Forming Asymmetric Responses in Enterprise Planning for Arms Development]," *Voennaya Mysl' [Military Thought]*, February 1, 2020, pp. 53–58.

[40] V. V. Selivanov and Y. D. Ilin, "Metodicheskiye Osnovi Formirovaniya Asimetricheskih Otvetov v Voyenno-Tekhnicheskom Protivoborstve s Visokoteknologichnim Protivnikom [Methodological Foundations for Forming Asymmetric Responses in Military-Technical Confrontation with a High-Technological Adversary]," *Voennaya Mysl' [Military Thought]*, February 1, 2019.

[41] Fadeev, 2019.

necessary military results, such as demoralizing the enemy, [and] causing him economic, political, and territorial damage without the obvious use of armed forces."[42] Russian author A. A. Shevcov, for example, claims that information confrontation has been carried out in Syria since the first days of the conflict in 2010. Numerous news agencies and other media outlets, he notes, have offered misleading, one-sided, or even false coverage of public protests, thus aiming to "impose the[ir] desired point of view."[43] This view is supported by reporting in the Russian media regarding the ongoing information war between Russia and the United States in Syria, with both sides accusing each other of killing civilians or of having links to the Islamic State.[44]

Key Strategic Documents

In this section, we analyze the key strategic documents that guide Russia's approach to information confrontation:

- National Security Strategy of the Russian Federation Until 2020
- Information Security Doctrine of the Russian Federation
- Basic Principles for the Russian Federation's State Policy in the Field of International Information Security to 2020
- Military Doctrine of the Russian Federation
- Conceptual Views on the Activities of the Armed Forces of the Russian Federation in the Information Space.

Table 3.1 provides an overview of the purpose and interrelationship of these documents. We structure our analysis in hierarchical order, starting with strategic guidance on overarching, national-level concerns and then turning to documents that specifically address the

---

[42] Fadeev, 2019.

[43] Shevcov, 2018, pp. 59–67.

[44] Alexey Naumov, "Russkii Otvet: Chevo Rossiiya Dobilas Za Chetire Goda Voini V Sirii [Russian Answer: What Russia Has Achieved in Four Years of War in Syria], *Lenta.ru*, September 30, 2019.

Table 3.1
Strategic Documents of the Russian Federation

Document Title	Year(s) Adopted	Purpose	Relationship to Other Strategic Documents
National Security Strategy of the Russian Federation (NSS)	2009, 2015	Defines the national interests and strategic national priorities of the Russian Federation, as well as goals, objectives, and measures in the field of domestic and foreign policy aimed at strengthening the national security of the Russian Federation and ensuring the long-term sustainable development of the country	Primary strategic document
Russian Military Doctrine (RMD)	2010, 2014	Presents the system of officially adopted views on preparation for armed defense of the Russian Federation	The RMD is based on the NSS and described in the NSS as defining the main provisions of military policy; focuses on and expands upon the military dimension of national security
Information Security Doctrine of the Russian Federation (ISD)	2016	Identifies strategic goals and main directions of information security	Based on the NSS, focuses on national security in the information domain
Basic Principles for the Russian Federation's State Policy in the Field of International Information Security to 2020	2013	Identifies the main threats to international information security; sets the goal, objectives, and priority policy directions in the information security domain; defines mechanisms for their implementation	Elaborates on the NSS and ISD (2000)
Conceptual Views on the Activities of the Armed Forces of the Russian Federation in the Information Space	2013	Defines the basic principles and rules guiding the armed forces of the Russian Federation in their use of the global information space to achieve defense and security objectives	Based on the ISD (2000) and RMD; elaborates on RMD provisions regarding the information domain

role of the armed forces of the Russian Federation in the sphere of information confrontation.[45]

National Security Strategy of the Russian Federation

The NSS, released in 2009, predicted that the "global information confrontation [would] intensify" as leading foreign states would seek to achieve "overwhelming superiority" though the employment of information warfare.[46] The most recent NSS, released in 2015, does not use the term "information confrontation."[47] Instead, it references the broader notion of "confrontation in the information sphere." Confrontation in the information sphere has become increasingly influential in international politics, according to the NSS, because the United States and its allies rely on "information and communication technologies to reach their geopolitical objectives," including the manipulation of public consciousness in Russia and the falsification of Russia's history.[48] The NSS also notes that Russian national security is vulnerable to information activities conducted in the economic, educational, health care, and cultural spheres. It emphasizes that information, communication, and other advanced technologies pose additional national security threats, having given rise to new forms of illegality, and are associated with transnational organized crime.

Although the NSS does not explicitly refer to or define information confrontation, it provides a foundation for understanding the concept. The NSS suggests that IPb constitutes a spectrum of activities,

---

[45] In recent decades, Russia has been quite productive in generating strategic documents of various types and forms. Russia's Law on Strategic Planning (2014), while it defines the purposes of various strategic documents, does not clarify the hierarchy or interrelation of these documents. Mikhail Mushinsky, "Strategii, Kontseptsii, Doktriny v Pravovoy Sisteme Rossiyskoy Federatsii: Problemy Statusa, Yuridicheskoy Tekhniki i Sootnosheniya Drug s Drugom [Strategies, Concepts, Doctrines in the Legal System of the Russian Federation: Problems of Status, Legal Technique and Interrelation]," *Yuridicheskaya Tekhnika [Legal Technique]*, No. 9, 2015.

[46] Russian Federation, *Strategiya Natsional'noy Bezopasnosti Rossiyskoy Federatsii do 2020 Goda [National Security Strategy of the Russian Federation Until 2020]*, May 2009.

[47] Russian Federation, *Strategiya Natsional'oi Bezopsnosti Rossiyskoy Federatsii do 2020 Goda [National Security Strategy of the Russian Federation Until 2020]*, December 2015.

[48] Russian Federation, *NSS*, 2015.

from psychological manipulation to the falsification or alteration of historical documents to criminal interference with information infrastructure. It suggests that in response to this spectrum of activities, Russia should construct a more robust information infrastructure and promote equal access to information resources across Russia's territory. While the NSS does not advocate limiting access to the information sphere, it does advocate for control of the information sphere to protect Russian citizens from external ideological influences. Finally, while the NSS does not directly refer to information as a weapon of war, it does characterize informational means as a tool that can be used to achieve strategic deterrence and preserve Russia's sovereignty and territorial integrity.

Information Security Doctrine of the Russian Federation

One of the general provisions of the NSS focuses on ensuring national security by concentrating national efforts and resources in, among other spheres, the information sphere. Like the NSS, the ISD of the Russian Federation does not explicitly refer to the concept of information confrontation, but it provides insight into Russian perceptions of threats in the information sphere.[49] The ISD equates threats to information security with threats to national security,[50] Russia's sovereignty, strategic stability, territorial integrity, and Russia's cultural and historical identity, suggesting that Russia fears the use of informational means by external forces to challenge the "sovereignty, political and social stability, [and] territorial integrity of the Russian Federation and its allies."[51] As information and information technologies become interwoven into every sphere of national life, the ISD recognizes that threats to information security are both technological

---

[49] Russian Federation, 2016.

[50] "National security" is defined as the state of protection of individuals, society, and government from internal and external threats, which allows for the provision of constitutional rights, freedom, sufficient quality and standards of living for citizens, sovereignty, territorial integrity, and the resilient development of the Russian Federation, defense, and security of the state. Russian Federation, NSS, 2015, p. 3; Russian Federation, ISD, 2016, p. 2.

[51] Andrei Shitov, "Dve Doktrini: Chem Otlichajutsa Podhodi Rasii u SSHA k Informacionnoi Bezopastnosti [Two Doctrines: What Are the Differences Between the Russian and U.S. Approaches to Information Security]," TASS, December 12, 2019; Russian Federation Information Security Doctrine, as translated in Hodgson et al., 2019.

and psychological in nature. Ensuring information security, therefore, requires "regulatory, intelligence, counterintelligence, operational research, scientific and technical, information analytical, economic, and other means to predict, detect, deter, prevent, and respond to information threats and to eliminate the consequences of their realization."[52]

This broad conception of threats to information security reflects the perception that the information space is vast, encompassing infrastructure, information producers and consumers, and internet, communication, and social media sites across Russia's territory, territories under Russia's influence, and territories that host elements of Russia's communications infrastructure in accordance with international agreements. While the NSS states that Russia should exert control over the information space, and in particular the Russian portion of the internet, the ISD recognizes the transnational nature of Russia's information space.[53] The ISD notes that technologically advantaged countries will necessarily dominate the information space. It calls for Russia to improve its information systems, including military systems and automated control systems. Whereas the NSS delineates a spectrum of information activities aimed at Russia, the ISD provides additional detail on the particular characteristics of the information space. It emphasizes the duality of information networks as both the target of a variety of threats and a means of responding to and defending against those threats.

Basic Principles for the Russian Federation's State Policy in the Field of International Information Security

As noted in the ISD, Russia's information security depends on international information security.[54] The *Basic Principles for the Russian Federation's State Policy in the Field of International Information Security to 2020* follows from the ISD's assertion that technologically advantaged countries will dominate the global information space. The *Basic Principles* refer to "information weapons," rather than IPb, but they pro-

---

[52] Russian Federation, 2016, p. 2.

[53] Russian Federation, 2016, p. 12.

[54] The Basic Principles defines "international information security" as the state of global information space where the following do not exist: the violation of individual, public, and state rights in the information sphere, as well as destructive and illegal impact on elements of national critical information infrastructure.

vide insight into how Russia views the global information space. The *Basic Principles* note that in today's society, "information and communication technologies are the primary factor that determines the level of socio-economic development and state of national security."[55] The *Basic Principles* prioritize the protection of Russia's information and communication technologies from threats, including the use of information and communication technologies as "information weapon[s] for military and political gain" that can be used to violate a state's sovereignty and territorial integrity and threaten international peace, security, and strategic stability.[56] Other threats to the global information space include acts of terrorism aimed at the security of Russia's critical information infrastructure, interference by foreign actors in the domestic issues of sovereign states, and computer crimes associated with obtaining illegal access to information.

Military Doctrine of the Russian Federation

In recent years, successive iterations of the RMD have noted the growing importance of information in modern warfare. Although the RMD has not offered a definition of IPb, it lists information activities as being among the major elements of military conflict. The 2010 RMD named the "increasing role of information confrontation" as a characteristic trait of modern military conflicts. It went on to identify states' employment of information campaigns in order to curry favor with international audiences as a core features of 21st-century competition.[57] Informational means could be used to "achieve political objectives without the use of military force and, subsequently, in the interest of shaping a favorable response from the global community to the use of military force." Perhaps most notably, this marked the first time that Russian doctrine had identified the "development of forces

---

[55] National Security Council of the Russian Federation, *Osnovy Gosudarstvennoy Politiki Rossiyskoy Federatsii v Oblasti Mezhdunarodnoy Informatsionnoy Bezopasnosti na Period do 2020 Goda [Basic Principles for the Russian Federation's State Policy in the Field of International Information Security to 2020]*, July 24, 2013.

[56] National Security Council of the Russian Federation, 2013, p. 2.

[57] Russian Federation, *Voyennaya Doktrina Rossiyskoy Federatsii [Military Doctrine of the Russian Federation]*, December 15, 2014b.

and means of information confrontation" as a priority task for equip-
ping Russian armed forces.[58] Russian defense officials subsequently con-
firmed the establishment of an "information operations force" (*voyska
informatsionnykh operatsiy*) that included a mandate for digital asymmetric confrontation with perceived adversaries.[59]

The most recent version of the RMD, released in 2014, suggests
that the main threats to Russia come from the information space. It
views the information space as on par with other military domains and
recognizes that threats to the Russian military increasingly come from
the information space.[60] Among the main tasks of Russia in deterring
military conflict are assessing and forecasting the "development of the
military-political situation at the global and regional levels . . . with
the use of modern technical means and information technologies" and
reducing the "risk of using information and communications technologies for military-political purposes."[61]

A "large-scale war against the Russian Federation becomes less
probable," the RMD notes, as external threats to Russia increasingly
come from the information space.[62] These external threats include the
"use of information and communication technologies for military-
political purposes to take actions that run counter to international law,
being aimed against the sovereignty, political independence, [and] territorial integrity of states and threatening the international peace, security, global, and regional stability," as well as "subversive information
activities against the population, especially the young citizens of the
state, aimed at undermining historical, spiritual, and patriotic tradi-
tions related to the defense of the Motherland."[63] The RMD characterizes the main elements of contemporary military conflict as the inte-

---

[58] Russian Federation, 2014b.

[59] "Istochnik v Minoborony: V Vooruzhennykh Silakh RF Sozdany Voiska Informatsionnykh Operatsiy [Source at the Ministry of Defense: The Armed Forces of the Russian Federation have Created Information Operations Forces]," *TASS*, May 12, 2014.

[60] Russian Federation, 2014b.

[61] Russian Federation, 2014b.

[62] Russian Federation, 2014b.

[63] Russian Federation, 2014b.

grated use of informational measures in combination with a "wide use of the protest potential of the population and special operations forces," while simultaneously exerting "pressure on the enemy throughout the enemy's territory in the global information space," as well as in other warfighting domains.[64] It further notes that information technologies can be used to assess and forecast changes in military-political relations at the global and regional levels.[65]

Conceptual Views on the Activities of the Armed Forces of the Russian Federation in the Information Space

The 2011 document *Conceptual Views on the Activities of the Armed Forces of the Russian Federation in the Information Space* characterizes information as a weapon that can be used against Russian interests, a target of hostile actors, and a means of protecting Russian interests. It treats information as a domain that is on par with the land, air, maritime, and cyberspace domains and delineates specific military activities, concepts, and systems that are associated with the information space.[66] The information space itself encompasses activities associated with the "formation, development, transformation, transmission, use, and storage of information which impacts individual and public consciousness, information infrastructure, and information itself."[67] The *Conceptual Views* does not define information confrontation. Instead, it defines "military conflict in the information space" as a "form of resolution of international or domestic tensions using information weapons" and "information weapons" as "information technology, means, and methods used to conduct information warfare."[68] It defines "information war" as follows:

> The confrontation between two or more states in the information space with the purpose of causing damage to information systems, processes and resources, critical and other infrastructure,

---

[64] Russian Federation, 2014b.

[65] Russian Federation, 2014b.

[66] Ministry of Defense of the Russian Federation, 2011.

[67] Ministry of Defense of the Russian Federation, 2011, p. 5.

[68] Ministry of Defense of the Russian Federation, 2011, p. 4.

undermining the political, economic and social systems, massive psychological manipulation of the population to destabilize the state and society, as well as coercing the state to make decisions in the interest of the opposing force.[69]

This definition omits the role of nonstate actors in the information space, although it links the technological and psychological aspects of information war.[70]

The *Conceptual Views* further note that the increasing use of the information space by the militaries of developed countries has made Russia's national information and communications systems more vulnerable to electronic attack, software manipulation over the internet, and digital manipulation through mass media platforms. Consistent with the observations made in the NSS and the ISD, the *Conceptual Views* affirm that the armed forces of the Russian Federation are prepared to defend Russian military and civilian information systems and infrastructure through the deterrence, prevention, and resolution of military conflicts in the information space by peaceful means. This mention of civilian systems underscores Russia's conviction that any and all threats to information security are also threats to national security.

Information Confrontation and International Law

The concept of information confrontation has important implications for Russian conceptions of international law. The Russian military-scientific and legal literature includes a variety of viewpoints on the role of international law as it pertains to information confrontation and information warfare. Some scholars emphasize the need to update existing international treaties and documents to include language that clearly reflects the realities of information confrontation in an infor-

---

[69] Ministry of Defense of the Russian Federation, 2011, p. 5.

[70] Jolanta Darczewska, "Russia's Armed Forces on the Information War Front: Strategic Documents," OSW (Center for Eastern Studies), No. 57, June 2016, p. 47.

matized society.[71] Other scholars go further and propose the creation of an international legal framework that would govern information confrontation.[72] This framework, according to Russian scholars, could include provisions on the nonproliferation of information weapons and restrictions on the use of propaganda and informational-psychological influence. Finally, some scholars argue that Russia must ensure that its information influence strategies are perceived as legitimate within the

[71] See P. I. Antonovich, 2011, pp. 43–47; I. N. Dylevskii, S. A. Komov, and A. N. Petrunin, "Ob Informatsionnykh Aspektakh Mezhdunarodno-Pravovogo Poniatiya 'Agressiya' [On Information Aspects of the International Legal Definition of 'Aggression']," *Voennaya Mysl' [Military Thought]*, No. 10, 2013; I. N. Dylevskii, V. O. Zapivakhin, S. A. Komov, S. V. Korotkov, and A. N. Petrunin, "Mezhdunarodniy Rezhim Nerasprostraneniya Informatsionnogo Oruzhiya: Utopiaya ili Real'nost'? [International Regime for Nonproliferation of Information Weapons: Utopia or Reality?]," *Voennaya Mysl' [Military Thought]*, Vol. 10, 2014; Nataliia Romashkina, "Informatsionnyy Suverenitet v Sovremennuyu Epokhu Strategicheskogo Protivoborstva [Information Sovereignty in the Contemporary Age of Strategic Confrontation]," *Informatsionnyye Voyny [Information Wars]*, Vol. 52. No. 4, 2019; Slipchenko, 2013.

[72] See Stanislav Chernichenko, "Ideologicheskaya Agressiya kak Primeneniye Sily v Mezhdunarodnom Prave [Ideological Aggression as the Use of Force in International Law]," *Ievraziyskiy Iuridicheskiy Zhurnal [Eurasian Law Journal]*, Vol. 128, No. 1, 2019; Margarita Doroshenko and Daniil Kazarin, "Mezhdunarodno-Pravovoye Regulirovaniye Deyatel'nosti SMI v Usloviyakh Informatsionnykh Voyn [International Legal Regulation of Media Activities in the Context of Information Wars]," *Molodoy Uchenyy [Young Scientist]*, No. 141, 2017, pp. 339–342; I. N. Dylevskii, V. O. Zapivakhin, S. A. Komov, S. V. Korotkov, and A. A. Krivchenko. "O Dialekte Sderzhivaniya i Predotvrasheniya Voyennykh Konfliktov v Informatsionnuyu Eru [On the Dialect of Deterrence and Prevention of Military Conflicts in the Information Era]," *Voennaya Mysl' [Military Thought]*, No. 7, 2016; N. I. Kostenko, "Mezhdunarodnaya Informatsionnaya Bezopasnost' v Ramkakh Mezhdunarodnogo Prava (Metodologiya, Teoriya) [International Information Security Within the Framework of International Law (Methodology, Theory)]," *Rossiyskiy Zhurnal Pravovykh Issledovaniy [Russian Journal of Legal Research]*, Vol. 5, No. 4 (17), 2018; S. A. Modestov, 2009; D. Sirotkin, A. Tyrtyshny, and A. Makarenkov, "Model' Pravovogo Regulirovaniya v Oblasti Informatsionnogo Protivoborstva [Model of Legal Regulation in the Field of Information Confrontation]," *Informatsionnyye Voyny [Information Wars]*, No. 3, 2016; Aleksandr Zinchenko and Anastasiia Tolstukhina, "Mir ili Voyna v Kiberprostranstve? [Peace or War in Cyberspace?]," *Mezhdunarodnaya Zhizn' [International Life]*, No. 9, 2018.

international community and propose using engagement in international law as a mechanism for achieving this legitimacy.[73]

The military-scientific literature emphasizes that the growing role of information in conflict means that the language used in existing international agreements aimed at preventing war and aggression is becoming obsolete and should be revised. In particular, international organizations and international legal frameworks must be revised to account for different levels of information confrontation. P. I. Antonovich, for example, notes the difference between the strategic level of information confrontation, which he characterizes as referring broadly to the informational aspects of geopolitical tensions between countries, and the operational level, which includes electronic warfare, cyberoperations, psychological influence. International legal frameworks should endeavor to regulate both the strategic and operational elements of information confrontation.[74]

Within the Russian military-scientific literature, there has been discussion of the concept of aggression in international law, and how this concept relates to information confrontation. Igor Dylevskii and colleagues, for example, address how the United Nations has historically understood and treated the concept of aggression, focusing in particular on gaps in United National General Assembly Resolution 3314, which was adopted in 1974. The Resolution describes aggression as the use of the armed forces of one country against the sovereignty, territorial integrity, or political independence of another country. This definition of aggression is based on a physical understanding of territory as comprising land, sea, and air, so it can only be applied to physical aspects of information confrontation, such as the destruction of the information and communication resources of an adversary or cyberattacks on its

---

[73] See Mikhail Kaverin, "Liderstvo Rossiyskoy Federatsii v Sisteme Mezhdunarodnykh Institutov [Leadership of the Russian Federation in the System of International Institutions]," *Informatsionnyye Voyny [Information Wars]*, Vol. 46, No. 2, 2018; Dmitriy Medvedev and Asiyat Tarchokova, "Instrumenty Mezhdunarodnoy Legitimatsii Vneshnepoliticheskikh Deystviy Rossii v Usloviyakh Informatsionnogo Protivoborstva [Instruments for the International Legitimation of Foreign Political Actions in Russia in the Conditions of Information Confrontation]," *Informatsionnyye Voyny [Information Wars]*, Vol. 47, No. 3, 2018.

[74] Antonovich, 2011.

infrastructure, where both the source and the target of the attack can be clearly identified.[75] Such attacks should be treated as violations of *information sovereignty*, defined by Nataliia Romashkina as the technological and legal capacity to defend a country's independence and the constitutional rights of its citizens from external threats, as well as the capacity to maintain control over the domestic information space.[76]

Dylevskii stresses, however, that under international law, only those attacks that are either conducted or directed by the armed forces of a specific country can be classified as legal aggression. Cyberattacks by nonstate actors, therefore, should be treated as criminal acts such as terrorism or extremism, but not as acts of aggression. The country that "hosts" such an attacker is not responsible for their actions unless a link between them and the armed forces or political leadership of that country is proven.[77]

If the first condition is met, such that the attacker entered the information space of a different country on behalf of its armed forces, the second condition, stated in the abovementioned resolution, is used to determine whether the attack is an act of aggression and, by extension, the seriousness of the consequences. If an act is identified as aggression, the country that coordinated the attack, as well as any countries that served as intermediaries (i.e., allowed the aggressor country to keep necessary servers on their territory) should be considered to be aggressors.[78]

Dylevskii and his coauthors propose that the United Nations should develop a separate set of criteria for identifying aggression in the information space.[79] Modestov, on the contrary, suggests that the Additional Protocol (I) to the Geneva Conventions, which prohibits certain types of weapons, may be applied to information weapons.[80]

---

[75] Dylevskii, Komov, and Petrunin, 2013.

[76] Romashkina, 2019.

[77] Dylevskii, Komov, and Petrunin, 2013

[78] Dylevskii, Komov, and Petrunin, 2013.

[79] Dylevskii, Komov, and Petrunin, 2013.

[80] Modestov, 2009.

The Russian military-scientific literature also characterizes informational-psychological confrontation, including propaganda and the incitement of radical views and actions, as a means of engaging in "ideological" aggression. Dylevskii identifies the export of democracy and democratic values through the media and information technologies by the West as an example of this type of information confrontation.[81] In a later article, Dylevskii notes that world leaders can use electronic media to quickly and easily disseminate messages that can drastically change public opinion, affect domestic politics and financial markets, and lead to the deterioration of bilateral relations, destabilization, civil unrest, and internal, regional, and international conflicts.[82]

According to Stanislav Chernichenko, however, ideological or psychological influence should not be equated with acts of aggression, as there is no direct use of military force or violation of territorial integrity. Nonetheless, he notes that information influence has a clear element of coercion and therefore poses a threat to international peace. Information influence can be used to prepare the leadership and population of an adversary for a future armed confrontation by internally destabilizing the enemy and laying the groundwork for the legitimization of further aggression.[83]

The literature identifies two approaches, international and regional, for developing a regulatory framework for information confrontation. The first approach relies on the regulation of information confrontation using international law. According to Dylevskii, the West believes that information confrontation should be treated similarly to other types of warfare, thereby giving victims of information aggression the right to defend themselves and collective defense organizations, such as NATO, a way to protect their member states. Russia, on the contrary, is advocating for the regulation of information confrontation as a separate form of aggression.[84] Sirotkin and colleagues empha-

---

[81] Dylevskii, Komov, and Petrunin, 2013.

[82] Dylevskii, Zapivakhin, Kovom, Korotkov, and Krivchenko, 2016.

[83] Chernichenko, 2019.

[84] Dylevskii, Zapivakhin, Kovom, Korotkov, and Krivchenko, 2016.

size the existence of a multipolar global system in which the United States, the European Union, China, and Russia are equally influential, and suggest that these influential powers should jointly develop a legal framework for information confrontation.[85] Dylevskii has proposed the construction of a nonproliferation regime for information weapons. He envisions this regime as consisting of a variety of bilateral and multilateral agreements.[86] According to Romashkina, Russia has already made substantial progress on this front, having concluded agreements with Brazil (2010), Belarus (2013), Cuba (2014), and China (2015).[87] The second approach relies on the regulation of information confrontation using regional legal frameworks. Here, too, the military-scientific literature stresses Russia's progress in partnering with its allies to address information confrontation, including through the Shanghai Cooperation Organization (2009), the Collective Security Treaty Organization (2010), and the Newly Independent States (2013).[88]

Russian scholars assign great importance to the development of an international framework for information security. Dylevskii has suggested the creation of a special United Nations body that would develop universally accepted rules and regulations surrounding information security. This body would distinguish between military and dual-use information weapons and impose limitations on their use and development.[89] According to Dylevskii, Russia has regularly proposed such limitations, but the West has opposed these efforts. Western opposition to Russian efforts to regulate the use of information weapons is perceived as the West "camouflaging its desire to further promote propaganda by appealing to freedom of speech and self-expression."[90] Aleksandr Zinchenko and Anastasiia Tolstukhina similarly characterize Russian efforts as a "peaceful plan," unlike the Western approach. They perceive

---

[85] Sirotkin, Tyrtyshny, and Makarenkov, 2016.

[86] Dylevskii, Zapivakhin, Kovom, Korotkov, and Petrunin, 2014.

[87] Romashkina, 2019.

[88] Romashkina, 2019.

[89] Dylevskii, Zapivakhin, Kovom, Korotkov, and Petrunin, 2014.

[90] Dylevskii, Zapivakhin, Kovom, Korotkov, and Krivchenko, 2016.

that the United States is balancing on the brink of war, being careful not to engage in outright aggression but diligently maintaining its superiority in information warfare. The United States will then leverage this superiority to control other countries. Zinchenko and Tolstukhina view the Budapest Convention, which they characterize as the only international agreement that pertains to information confrontation, as evidence of this trend, because Article 32(b) unacceptably paves the way for more powerful countries to intrude on the information space of weaker countries without consent. Russia, on the other hand, by calling for responsible behavior in the information space, endeavors to ensure equality between states and promote a semblance of global justice.[91]

The military-scientific literature also addresses the right to self-defense against information aggression. According to Chernichenko, the victims of information aggression have limited legal recourse, so they often must rely on making largely symbolic public statements.[92] Doroshenko and Kazarin similarly note that victims of information aggression are often limited to political statements, such as severing diplomatic relations and ceasing official contacts with the aggressor.[93] For a country to act in self-defense, Chernichenko explains, the U.N. Security Council should make a determination that the initial act of information influence was in fact an act of aggression and that the ideological influence was sufficient to give the victimized country a right to react.[94]

Kostenko proposes the creation of an entirely new field of international law for international information security. This field would address acts of aggression against the national sovereignty and territorial integrity of independent states, threats to international peace and stability within the information domain, intervention into the domes-

---

[91] Zinchenko and Tolstukhina, 2018.

[92] Chernichenko, 2019.

[93] Margarita Doroshenko and Daniil Kazarin, 2017.

[94] Chernichenko, 2019.

tic affairs of other countries, incitement, the use of cyber warfare, and information wars.[95]

The literature emphasizes, moreover, that while Russia pushes for the development of an international legal framework for information confrontation, it should also use informational means to improve its standing in the international community.[96] While Russia uses multilateral institutions such as the Collective Security Treaty Organization, the Eurasian Economic Union, the Customs Union, the Shanghai Cooperation Organization, and BRICS (Brazil, Russia, India, China, and South Africa) to promote its values and interests, it has not used the platforms offered by these institutions to their fullest extent to convince the international community of the legitimacy of Russian actions. Dmitriy Medvedev and Asiyat Tarchokova refer to three recent examples of such a failure: the cases of Georgia, Ukraine, and Syria. They note that Russia should have used its informational influence and geopolitical status to convince the international community of the legitimacy of Russian actions to avoid the imposition of sanctions and other retaliatory measures.[97]

---

[95] Kostenko, 2018.

[96] Kaverin, 2018.

[97] Medvedev and Tarchokova, 2018.

CHAPTER FOUR

Information Confrontation Through the Ukrainian Lens

In 2017, O. V. Levchenko, the director of Ukraine's Zhytomyr Military Institute, characterized Russian operations in Crimea and eastern Ukraine as the "most complete and illustrative example of a new generation war," a key component of which is information confrontation.[1] As a result, these operations—notably Russia's employment of informational instruments in Ukraine—have been the subject of many analyses and scholarly discussions. In the West, however, few observers have explored the rich discourse that exists within the Ukrainian military-scientific community on the topic of Russian information confrontation campaigns targeting Ukraine. Accordingly, this chapter examines Russian information confrontation through a new lens—that is, from the perspective of the target state's scholarly community. In the coming months and years, there will undoubtedly be new insights regarding Russian information confrontation in Ukraine.

As a Soviet republic before 1991, Ukraine and its military institutions were a constituent part of the broader Soviet defense establishment, which included a training and education system for military personnel. With the disintegration of the Soviet Union and Ukrainian independence, Kyiv inherited Soviet arms and equipment, a cadre of military personnel trained in Soviet institutions steeped in Soviet doctrine, and "34 [Soviet] military schools and faculties at 78 institutions of higher

---

[1] O. V. Levchenko, "Evolutsia Gibridnoi Vijiny Rosiys'koi Federatsii proty Ukraini [The Evolution of Russia's Hybrid War Against Ukraine]," *Nauka i Oborona [Science and Defense],* No. 2, 2017, p. 11.

learning."[2] Though the Ukrainian armed forces and associated military institutions have undergone several periods of reform, vestiges of their Soviet past remain.[3] For instance, Ukrainian military doctrine continues to reflect some hallmarks of Soviet doctrinal convention.[4] Personnel at the upper echelons of Ukrainian military institutions, in particular, are products of Soviet training and education, while younger generations are influenced by the "consanguinity" of the Ukrainian and Russian military systems as decedents of the same Soviet parent structure.[5] As a result, the Ukrainian Ministry of Defense has had to grapple with the challenge of Russian loyalists filling the senior ranks of the country's armed forces and the defense ministry itself. At one point, the leadership of the Ukrainian Ministry of Defense claimed that 40 percent of the organization's senior leaders failed polygraph tests designed to evaluate their allegiances.[6] Like their Russian counterparts, moreover, the Ukrainian defense community has carried on the tradition of cultivating a robust and prolific military-scientific community. In Ukraine, there is also a robust community of senior civilian scholars who work on issues related to military strategy and doctrine. The shared lineage of the Ukrainian and Russian defense establishments means that this discussion of the Ukrainian perspective on IPb serves as a useful foil to

---

[2] Sergiy Gerasymchuk, *Research Report on the Second Investigation Level: Ukrainian Case*, Frankfurt, Germany: Peace Research Institute Frankfurt, 2008, p. 2; James J. Tritten, *Military Doctrine and Strategy in the Former Soviet Union: Implications for the Navy*, Monterey, Calif.: Naval Postgraduate School, 1993, p. 78.

[3] For a discussion of the lasting influence of the Soviet armed forces on the Ukrainian security sector, see Olga Oliker, Lynn E. Davis, Keith Crane, Andrew Radin, Celeste Gventer, Susanne Sondergaard, James T. Quinlivan, Stephan B. Seabrook, Jacopo Bellasio, Bryan Frederick, Andriy Bega, and Jakub P. Hlavka, *Security Sector Reform in Ukraine*, Santa Monica, Calif.: RAND Corporation, RR-1475-1-UIA, 2016.

[4] Stuart Gallagher, "Mission Ukraine: U.S. Army Leads Multinational Training Group to Counter Russian Threat," *Association of the United States Army*, May 19, 2020; Anna Nemtsova, "Why Is Ukraine's War So Bloody? The Soviet Union Trained Both Sides," *The Daily Beast*, April 14, 2017.

[5] Isabelle Facon, *Reforming Ukrainian Defense: No Shortage of Challenges*, Paris: IFRI, No. 101, 2017, p. 14.

[6] Facon, 2017, p. 14, citing "Ukraine's Defense Minister: 40% of Ministry's Officials Fail Polygraph Tests," *UNIAN Information Agency*, April 25, 2016.

our analysis, in the preceding chapters, of IPb through the eyes of Russia's military-scientific community. Further, Ukraine's position as both a frequent target of Russian IPb campaigns and a proving ground for new informational weapons has provided its scholarly community with ringside seats to observe Russian IPb operations.[7]

This chapter first provides a broad overview of the Ukrainian military-scientific and civilian literature on information confrontation, highlighting key lexical debates and core themes that are characteristic of this discourse. Next, it outlines the broad contours of the Ukrainian military-scientific community's discourse on Russian IPb in Ukraine in the post-Soviet period, particularly since Russia's 2014 annexation of Crimea, from the perspective of the Ukrainian military-scientific community. Throughout, we compare the thematic elements of this scholarship to its Russian equivalent—that is, the Russian military-scientific community's views on information confrontation campaigns in Ukraine.

Information Confrontation in the Ukrainian Military-Scientific Literature

Terminology and Lexical Debates

Since 2014, Ukraine has been beleaguered by conflict. Ukraine has also been the object of numerous soft power efforts, including IPb campaigns carried out by actors suspected to be affiliated with Russia.[8] These campaigns, often classified as hybrid war or new generation war, have shaped Ukrainian politics, policies, and viewpoints.[9] The Ukrainian military-scientific literature principally discusses Russian IPb directed

---

[7] See Mariia Zhdanova and Dariya Orlova, *Computational Propaganda in Ukraine: Caught Between External Threats and Internal Challenges*, Working Paper 2017.9, Computational Propaganda Research Project, Oxford University, 2017.

[8] Zhdanova and Orlova, 2017.

[9] It is important that we acknowledge that a central conceptual debate exists within the scholarship examining events that unfolded in Eastern Ukraine since 2014. Some argue the events should be deemed a civil war whereas others argue it should be referred to as an inter-state conflict. For a discussion of this debate, see Taras Kuzio and Paul D'Anieri, *The Sources*

at Ukraine in the context of hybrid warfare.[10] Ukrainian military scholars classify information confrontation as one of several instruments that states can employ in the conduct of hybrid warfare campaigns.

Broadly speaking, Ukrainian military and civilian experts on information confrontation in hybrid warfare use similar nomenclature. The terms information warfare (*informacijna vijna*), information confrontation (*informacijne protyborstvo*), and information operations (*informacijna operaciya*) are most commonly used to describe Russian activities in Ukraine.[11] That said, there appears to be little conceptual consistency across the military-scientific literature, with conceptual constructs varying from author to author. Some scholars use these related terms interchangeably. For example, in an article on the role of information warfare in international relations, M. O. Kondratyuk conflates the terms information warfare and information confrontation.[12]

By contrast, a Security Service of Ukraine (SBU) textbook on the history of informational-psychological confrontation from 2012 employs both information warfare and information confrontation,[13] while characterizing actions of informational influence (*akciyi informacijnoho vplyvu*), or AIV, special information operations (*special'na informacijna operaciya*), or CIO, and information warfare as different forms of information confrontation.[14] As in the Russian military-

---

of Russia's Great Power Politics: Ukraine and the Challenge to the European Order, Bristol, U.K.: E-International Relations Publishing, June 25, 2018, pp. 86–92.

[10] For examples of Ukrainian military-scientific literature that discusses information confrontation in the context of hybrid warfare, see V. P. Bocharnikov and S. V. Svyeshnikov, "Pohlyady na Kharakter Suchasnykh Voyennykh Konfliktiv [Views on the Character of Modern Military Conflicts]," *Nauka i Oborona [Science and Defense]*, No. 1, 2017.

[11] Because this chapter focuses on Ukrainian scholarship on and perceptions of IPb, here we use Ukrainian transliterations of key terms rather than Russian transliterations.

[12] M. O. Kondratyuk, "Informacijna Vijna ta Rol Mas-Media v Mizhnarodnyh Konfliktah [Information War and the Role of Mass Media in International Relations]," *Social'ni Komunikaciyi [Social Communications]*, Chapter II, Vistnyk [Journal], No. 41, 2013.

[13] Y. Zharkov, ed., *Istoriia Informatsiino-Psykholohichnoho Protyborstva [History of Information and Psychological Confrontation]*, Kyiv, Ukraine: Research and Publishing Department of the National Academy of Security Service of Ukraine, 2012, p. 147.

[14] Zharkov, 2012, p. 20.

scientific literature, the SBU textbook identifies broad and narrow types of IPb.[15] Broad IPb refers to a variety of military, diplomatic, economic, and other activities intended to influence the information domain of one's adversary while protecting one's own information space. By contrast, narrow IPb consists of actions that exploit information in order to achieve information superiority in a military context. The SBU textbook draws an important conceptual distinction between information war and information confrontation. It notes that while information wars typically include actions of informational influence and special information operations, these activities can also be carried out independently and outside the context of information warfare.

The SBU textbook cites the U.S. Army's doctrinal definition for information warfare from the mid-1990s. The textbook explains that in the 1996 iteration of *FM 100-6 Information Operations*, information warfare is defined as "actions taken to achieve information superiority by affecting adversary information, information-based processes, information systems, and computer-based networks while defending one's own information, information-based processes, information systems and computer-based networks."[16] By 2012, when the SBU textbook was published, the U.S. Army had released several updates to its information operations doctrine, but the SBU textbook does not reference those updates. Even so, the SBU textbook reflects a broader trend shared by the Russian and Ukrainian military-scientific communities—the tendency to point to the United States as the principal architect of the concepts of information warfare and information confrontation.[17]

---

[15] Zharkov, 2012, pp. 26–27.

[16] U.S. Department of the Army, Headquarters, *FM 100-6 Information Operations*, Washington, D.C., August 1996, p. 2-2. Interestingly, the authors of the SBU text choose to not include the field manual's definition of its namesake concept, information operations, and instead focus on information warfare.

[17] For examples in the Ukrainian literature, see Kondratyuk, 2013, p. 111; V. A. Lipkan, "Suchasnyj Zmist Informacijnyh Operacij Proty Ukrayiny [An Accounting of Current Information Operations Against Ukraine]," *Aktual'ni Problemy Mizhnarodnyh Vidnosyn [Current Issues of International Relations]*, Pub. 102, Sec. 1, 2011. For examples in the Russian literature, see Rodionov, 1998, pp. 67–70.

Elsewhere in the Ukrainian military-scientific literature, the distinction between information warfare, information confrontation, and other related terms is described according to alternative constructs. V. A. Lipkan, for instance, argues that the differences between information warfare, information confrontation, and information struggle (*informacijna borot'ba*) are "all manifestations of a broader concept— threats to national interests and national security in the information sphere."[18] Lipkan proposes a typology that distinguishes between these related concepts. Information confrontation, Lipkan argues, is a "rivalry of social systems," whether these systems are states or blocs of states, and occurs in the information sphere.[19] Lipkan explains that states engage in IPb to establish control over vital strategic resources. He identifies three types of information confrontation—information expansion, information aggression, and information warfare—which are distinguished by the scope of the effort, the intensity of the effort, and the means used to conduct the effort. Of these three types, information expansion is the least bellicose. States engage in information expansion when they use noncombative means to penetrate an adversary's information environment for their own benefit. On the other hand, when states perform targeted activities designed to inflict "specific, tangible damage" on their adversaries through the "limited and local application of information influence," they are conducting information aggression.[20] Lastly, Lipkan argues, information warfare is characterized by its broader scope and extent of hostility. It is, in the words of Lipkan, the "highest form of information confrontation conducted," involving the conduct of "information violence" (*informacijnoho nasyl'stva*) on a massive scale.[21] It is used to cause major upheavals in the international system, such as the ousting of sitting governments as a means of resolving major rivalries between actors.

These are far from the only constructs that exist in the Ukrainian literature on information confrontation. Some Ukrainian experts, such

---

[18] Lipkan, 2011, p. 37.

[19] Lipkan, 2011, p. 37.

[20] Lipkan, 2011, p. 38.

[21] Lipkan, 2011, p. 38.

as Kondratyuk, assert that information warfare refers to the integration of two types of struggle—informational-technical struggle and informational-psychological struggle—in the information sphere.[22] This construct mirrors the prevailing typology of information confrontation in the Russian military-scientific literature. That said, Kondratyuk's use of information warfare as the umbrella term for these two types differs from the Russian military-scientific literature, which tends to use the term information confrontation when referring to informational-technical and informational-psychological efforts and effects.

The Ukrainian military-scientific literature introduces several alternative terms into the scholarly lexicon. O. A. Zavodovs'ka, for example, refers to both the "information component of hybrid war" and "information confrontation," using these terms interchangeably.[23] I. Yu. Yuzova, the acting chief of the Sergeants' College at Ukraine's Ivan Kozhedub National Air Force University, defines "informational-psychological operations" (*informacijno-psyholohichni operaciyi*, or IPO), as "a system of information acts, attacks, and actions that are internally coherent and interconnected by purpose, tasks, objects, and time and are conducted simultaneously or sequentially under a unified plan for a joint purpose of informational-psychological influence on the target audience."[24] As might be expected, informational-psychological influence (*informacijno-psyholohichnyj vplyv*) is conceived of as the use of informational tools to influence individuals' or groups' attitudes and behaviors. Yuzova borrows this concept from a 2006 Ukrainian Min-

---

[22] Kondratyuk, 2013.

[23] O. A. Zavodovs'ka, "Formuvannya Informacijnoho 'Poryadku Dennoho' Yak Metod Vedennya Hibrydnoyi Vijny v Konteksti Suchasnyh Mizhnarodnyh Vidnosyn [Formation of the 'Daily Agenda' as a Method of Conducting Hybrid Warfare in the Context of Contemporary International Relations]," *Visnyk [Journal] of the Lviv University, International Relations Series*, Vol. 36, No. 3, 2015, pp. 32–39.

[24] I. Yuzova, "Analiz Orhanizatsiyi ta Vedennya Informatsiyno-Psykholohichnykh Operatsiy Pry Vedenni Hibrydnoyi Viyny [Analysis of the Organization and Conduct of Informational-Psychological Operations in the Conduct of Hybrid Warfare]," *Zbirnyk Naukovykh Prats' Kharkivs'koho Natsional'noho Universytetu Povitryanykh Syl [Anthology of Research Works of Kharkiv National Air Force University]*, No. 2, 2020, p. 41.

istry of Defense textbook on the theory and practice of informational-psychological operations.[25]

Ukrainian Perspectives on the Theory and Practice of Information Confrontation

The Ukrainian military-scientific literature acknowledges the profound impact of the information age on interstate relations and the conduct of warfare. The 2012 SBU textbook, for example, notes that "confrontation in the information sphere has become so significant that it has resulted in the necessary establishment of special concepts, which have been termed 'information warfare' or 'information confrontation.'"[26]

Though a number of Ukrainian military experts acknowledge that states have used informational tools since the dawn of warfare itself, they also note that information campaigns have evolved in recent years. The emergence of new technologies that facilitate the communication, consumption, and storage of information has shaped how states use and rely on information, which has in turn influenced the doctrinal development and implementation of information confrontation. In the SBU textbook, for example, M. M. Prysiazhniuk, describing the evolution of IPb, notes that state-sponsored information campaigns, which were once conceived of as "an aid in solving combat tasks at the tactical level," have evolved and achieved an elevated status as "a global function of managing [the] political process at a strategic level."[27] In short, information campaigns once played supporting roles in combat, but they have become a centerpiece of modern warfare.[28]

In this vein, a number of Ukrainian scholars have come to the conclusion that *all* modern military actions have an informational component.[29] Zavodovs'ka, for instance, asserts that information cam-

---

[25] Yuzova, 2020, p. 41.

[26] Prysiazhniuk, in Zharkov, 2012, p. 147.

[27] Prysiazhniuk, in Zharkov, 2012, p. 140.

[28] Prysiazhniuk, in Zharkov, 2012, p. 154.

[29] See Lipkan, 2011.

paigns are the "main feature of hybrid war" today.[30] Some scholars, like Yuzova, go so far as to say that today, global powers engage in information campaigns in advance of *all* kinetic military activities in the hope that degrading an adversary's information sphere prior to hostilities will improve their own prospects for victory.[31] Similarly, Ye. A. Makarenko, a Ukrainian professor of international relations, explains that states that are able to mount IPb campaigns first will obtain an informational advantage and therefore be more likely to prevail.[32] On the psychological side of IPb, Zavodovs'ka explains that states use the psychological component of IPb to lay the groundwork for future operations well in advance. They do so through the use of "special agents," as well as local media, that are tasked with propagating narratives likely to curry favor with local audiences while maligning local authorities.[33]

In contrast with earlier periods, when states inflicted physical destruction on the battlefield in the service of political objectives, today's next generation wars may be principally conducted "through battles on the information field," which Ukrainian scholars have characterized as being inherently less bloody than their conventional counterparts.[34] By leveraging emerging information technologies, modern militaries can optimize their operations and maximize their efficiency and effectiveness without inflicting—or suffering—heavy losses, especially in comparison with traditional forms of warfare.[35]

Although this suggests that conflicts in the information age may result in fewer casualties, this does not mean that informational weapons are any less pernicious than conventional weapons. Ukrainian military experts draw parallels between information confrontation campaigns (in their many terminological permutations) and weapons of

---

[30] Zavodovs'ka, 2015, p. 33.

[31] Yuzova, 2020, p. 41; Zavodovs'ka, 2015, p. 34.

[32] Ye. A. Makarenko, "Informacijne Protyborstvo u Suchasnyh Mizhnarodnyh Vidnosynah [Information Confrontation in Modern International Relations]," *Mizhnarodni Vidnosyny, Seriya «Polityčhni Nauky» [International Relations, Political Science Series]*, No. 17, 2017.

[33] Zavodovs'ka, 2015, p. 34.

[34] Yuzova, 2020, p. 41; Makarenko, 2017.

[35] Makarenko, 2017.

mass destruction (WMD).[36] When wielded as instruments of influence, both IPb and WMD have the potential to reshape the international system.[37] In this respect, the Ukrainian military-scientific literature parallels the Russian literature, which also likens IPb weapons to WMD.[38] As V. A. Lipkan notes, moreover, in the absence of established legal norms relating to information confrontation, states can conduct highly destructive informational activities in the service of political goals but suffer few concrete repercussions.[39] The Ukrainian military-scientific literature characterized a state's ability to wield information as an equalizer, enabling states with inferior firepower to achieve an informational advantage and prevail over adversaries with greater conventional capabilities.[40] This suggests, according to Yuzova, that the carnage associated with large-scale conventional wars may become a thing of the past.[41]

Much like information confrontation itself, the Ukrainian military-scientific literature on IPb has also evolved. As might be expected, Russia's annexation of Crimea and incursions into eastern Ukraine in 2014 provoked a wave of analysis focused on IPb, particularly in the context of hybrid warfare.[42]

---

[36] Yuzova, 2020.

[37] Yuzova, 2020; Vitalij Mykolajovych Butuzov and Kateryna Viktorivna Titunina, "Suchasni Zahrozy: Komp'yuternyj Teroryzm [Current Threats: Computer Terrorism]," *Borot'ba z Orhanizovanoyu Zlochynnistyu i Korupciyeyu [Fight Against Organized Crime and Corruption]*, No. 17, 2007, pp. 316–324; Makarenko, 2017.

[38] Yu. Sinchuk, "Sposoby Vedenija Sovremennyh Vojn [Methods of Conducting Modern Wars]," *Voennaya Mysl' [Military Thought]*, 2000.

[39] Lipkan, 2011, p. 38.

[40] Zavodovs'ka, 2015, p. 36

[41] Recent hostilities in Ukraine suggest, however, that such carnage may not be avoided in all cases.

[42] Levchenko, 2017. Additionally, Yuzova (2020) cites a number of studies advancing the theoretical development of the concepts of both hybrid warfare as well as information confrontation. These include Volodymyr Horbulin, *Svitova Hibrydna Viina: Ukrainskyi Front [The World Hybrid War: Ukrainian Forefront]*, Kyiv, Ukraine: National Institute for Strategic Studies, 2017; Fedir Turchenko and Halyna Turchenko, *Proiekt "Novorosiia" and Novitnia Rossiisko-Ukrainska Viyna [The Novorossiya Project and the Latest Russian-Ukrainian War]*, Kyiv, Ukraine: Institute

The View from Kyiv: Russia's IPb in Ukraine, 2014 to Present

As in Russia, Ukrainian conceptions of IPb have been informed by a handful of post–Cold War conflicts, principally the Gulf War, the Kosovo War, the Chechen Wars, and the Russo–Georgian War.[43] That said, the Russian and Ukrainian military-scientific literatures diverge in their interpretations of Russia's employment of information confrontation as a tool of influence. Nowhere is this more evident than in examinations of information campaigns targeting Ukraine, particularly those associated with Russian military operations in Ukraine since 2014.

Ukrainian military experts have spilled considerable ink on the topic of Russian IPb campaigns in Ukraine since 2014. The Ukrainian military-scientific literature unequivocally names Russia as the culprit behind recent information warfare campaigns targeting Ukraine.[44] Ukrainian scholars offer additional detail on Russia's suspected motiva-

---

of History of Ukraine, 2015; D. Prysiazhniuk, "Zastosuvannya Manipuliatyvnykh Psykho-tekhnolohii z Boku Rosii v ZMI Ukrainy (Na Prykladi Krymu) [Application of Manipula-tive Psychotechnologies by Russia in Ukrainian Media (on the Example of Crimea)]," *Visnyk Kyyivs'koho Natsional'noho Universytetu Imeni Tarasa Shevchenka: Viys'kovo-Spetsial'ni Nauky [Journal of the Taras Shevchenko National University of Kyiv: Military-Special Sciences]*, No. 23, 2009, pp. 63–66; H. Pievtsov, A. M. Hordiienko, S. V. Zalkin, S. O. Sidchenko, A. O. Feklistov, and K. I. Khudarkovskyi, *Informatsiino-Psykholohichna Borotba u Voien-nii Sferi [Information and Psychological Warfare in the Military Domain]*," Kharkiv, Ukraine: Kharkiv National University, 2017; H. Yavorska, "Hibrydna Viina yak Dyskursyvnyi Kon-strukt [Hybrid Warfare as a Discursive Construct]," *Stratehichni Priorytety [Strategic Priori-ties]*, No. 4, 2016, pp. 41–48; M. Dziuba, Ya. M. Zharkov, I. O. Olkhovoi, and M. I. Onysh-chuck, *Narys Teorii i Praktyky Informatsiino—Psykholohichnykh Operatsii [Essay on the Theory and Practice of Information and Psychological Operations]*, Kyiv, Ukraine: VITI NTUU "KPI," 2006; Pavlo Hai-Nyzhnyk, ed., *Ahresiia Rosii Proty Ukrainy: Istorychni Peredumovy ta Suchasni Vyklyky [Russia's Aggression Against Ukraine: Historical Background and Current Challenges]*, Kyiv, Ukraine: MP Lesya, 2016.

[43] Prysiazhniuk, in Zharkov, et al., 2012, p. 8.

[44] O. Yu. Ivanov, "Rosijs'ko-Ukrains'ke Informacijne Protiborstvo z «Krims'kogo Pitannia»: Genezis Ta Suchasnij Stan [Russian-Ukrainian Confrontation and the 'Crimea Question': Genesis and Current Status]," in *Current Problems Related to the Management of Information Security of the Government*, 8th Scientific and Applied Conference, Kyiv, Ukraine: Ukrai-nian Ministry of Education and Science, Institute for the Modernization of Educational Content of Ukraine, National Academy of the Security Service of Ukraine, and the Research

tions in unleashing information campaigns targeting Ukraine. Although there are differences between Ukrainian scholars in their characterization of Russian aims, they generally converge on the objectives of Russian IPb in Ukraine. These objectives are to

- demoralize the Ukrainian public, its armed forces, and its security services,
- seed new and expand existing divisions within the Ukrainian public, which can be exploited in future operations,[45]
- mischaracterize events such that they are favorable to Russia,[46]
- establish support for Russian actions in Ukraine among Ukrainian audiences,[47]
- reassert Russian influence over Kyiv authorities,[48]
- foster divisions among the Ukrainian population, creating a favorable ground for future operations.[49]

Some Ukrainian military experts suggest that the seeds of the active phase of the Kremlin's ongoing information operations in Ukraine were planted years earlier.[50] According to the findings of an international roundtable hosted by Ukraine's Institute of International Relations at the Taras Shevchenko National University of Kyiv, Russia's preparatory activities began as early as 2006 or 2007 (in the years following the Orange Revolution in Ukraine).[51] It was at this time,

Institute of Informatics and Law of the National Academy of Legal Sciences of Ukraine, 2017, pp. 38–40.

[45] Levchenko, 2017, p. 13.

[46] Levchenko, 2017, pp. 11–13.

[47] "Politychni Komunikaciyi za Umov Mizhderzhavnyh Konfliktiv [Political Communications for the Purpose of Interstate Conflict]," *Analytical Report Based on the Results of an International Roundtable*, Kyiv, Ukraine: Taras Shevchenko University of Kyiv, Institute of International Relations, March 30, 2015, p. 3.

[48] Levchenko, 2017, p. 11.

[49] Levchenko, 2017, p. 13.

[50] "Politychni Komunikaciyi," 2015, p. 3; Ivanov, 2017, pp. 38–40.

[51] "Politychni Komunikaciyi," 2015, p. 3.

scholars suggest, that Russian activities in the Ukrainian information space began to exhibit attributes characteristic of a coordinated, deliberate campaign—that is, according to roundtable participants, their methods, goals, and timing became more systematic.[52]

Zavodovs'ka similarly asserts that information wars, such as those conducted by Russia in Ukraine, require fertile ground to be successful. In advance of information campaigns, actors like Russia will cultivate the target population using informational tools.[53] Levchenko, for example, cites a famous 2013 speech by Gerasimov at the Russian Academy of Military Sciences as evidence that Russian plans to conduct hybrid warfare in Ukraine predated its annexation of Crimea in 2014.[54] Gerasimov notes that in modern wars, "the emphasis has shifted to the use of political, economic, informational, humanitarian, and other non-military measures along with the use of the protest potential of the local population." He goes on to say that these soft power instruments "must be accompanied by covert military operations— for example methods of information warfare and the use of special forces."[55] Based on this, Levchenko divides Russian hybrid warfare in Ukraine into three phases: the preparatory phase, the active phase, and the consolidation phase.[56]

O. Yu. Ivanov of the National Academy of the SBU divides Russian hybrid warfare in Ukraine into two distinct phases.[57] According to Ivanov, the first phase began in the 1990s, when Russian officials sought to protest the newly independent Ukrainian government's territorial rights to Crimea. At that time, Russian officials disseminated messaging designed to challenge the legality of the Soviet decision to cede oversight of Crimea to the Ukrainian Soviet Socialist Republic in 1954. The Kremlin did so, Ivanov asserts, with assistance from Crimean

---

[52] "Politychni Komunikaciyi," 2015, p. 3.

[53] Zavodovs'ka, 2015, p. 36

[54] Levchenko, 2017, p. 11–12.

[55] V. Gerasimov, "Cennost' Nauki v Predvidenii [The Value of Science in Foresight]," *Voenno-Promyshlennii Kur'er [Military-Industrial Courier]*, 2013.

[56] Levchenko, 2017, p. 12.

[57] Ivanov, 2017, pp. 38–40.

politician Yuri Meshkov, who used the media to propagate narratives that called into question Crimean cultural, economic, and historical ties to Ukraine while affirming Crimea's ties to Moscow. In the late 1990s, Ivanov notes, there was a wave of historical writings that traced the Crimean lineage to the Russian tsarist period and beyond, which serves as evidence of the origins of Russian objectives and information efforts in Ukraine. According to Ivanov, this messaging campaign appears to have succeeded, with separatist sentiments in Crimea becoming more prominent in this period.[58] Ivanov identifies the second phase as commencing in 2014, after Moscow's annexation of the Crimean peninsula. In this period, Russian messaging propagated narratives championing the economic and social benefits of realignment with Russia. These narratives were not only directed at Crimean, broader Ukrainian, and international audiences, but also at the Russian public.[59]

Levchenko observes a similar trend in his analysis on the evolution of Russian hybrid warfare in Ukraine. The Kremlin's years-long employment of soft power instruments in Ukraine, including informational campaigns, cultivated circles of Russian sympathizers in Ukraine who were embittered by the authorities in Kyiv. Throughout the post–Cold War era, Levchenko argues, Moscow has propagated messaging in Ukraine designed to celebrate Russian achievements in science and technology, extol the prowess and heroism of the Russian armed forces, trumpet the allegedly higher standards of living in Russia, and affirm the shared historical, cultural, and territorial links between Ukraine and Russia.[60] By the same token, Levchenko notes, Ukrainians have been exposed to narratives belittling Ukraine's role in regional history and deliberately misconstruing historical events. Ultimately, these informational efforts have led to the creation of a so-called fifth column within the Ukrainian populace,[61] notably in

---

[58] Ivanov, 2017, pp. 38–40.

[59] Ivanov, 2017, pp. 38–40.

[60] Levchenko, 2017, p. 13.

[61] In his article on Russian hybrid warfare, Alexander Lanoszka defines fifth columns as "groups of individuals, usually acting covertly, embedded within a much larger population that they seek to undermine. Fifth columns may agitate or may simply wait for hostilities to

the southeastern regions of Ukraine, that was activated during the events of 2014.[62]

According to the Ukrainian military-scientific literature, Russian IPb campaigns in Ukraine use a variety of different forums to promote this messaging, including both traditional and digital media, events, and in-person interactions. Yu. V. Kovtun of the SBU details several of these tools in his article on the threat posed by Russian aggression in the Ukrainian information sphere.[63] The Russian security services, he argues, manage groups masquerading as organically arising Ukrainian groups that are displeased with Ukrainian government policies on social media platforms, including Vkontakte, Facebook, and Odnoklassniki. Russia also controls online forums, blogs, and other internet sites that mimic Ukraine's ".ua" domain name. Similar tactics include the deliberate use of Ukrainian words in website titles to feign a Ukrainian identity or affiliation.[64] According to Kovtun, these sites, which appear as though they are hosted in Ukraine, propagate anti-Ukrainian

---

break out between the target and the belligerent before becoming active. Such fifth columns may facilitate the next military campaign of the government they support at an opportune moment." See Alexander Lanoszka, "Russian Hybrid Warfare and Extended Deterrence in Eastern Europe," *International Affairs*, Vol. 92, No. 1, 2016, p. 179.

[62] Levchenko, 2017, p. 14. Other Ukrainian scholars have addressed the role of historical narratives as part of Russian IPb efforts in Ukraine. See, for example, M. S. Romanov, "The Participation of the Russian Federation's Scientific and Educational Institutions in Special Information Operations," in *Current Problems Related to the Management of Information Security of the Government*, 8th Scientific and Applied Conference, Kyiv, Ukraine: Ukrainian Ministry of Education and Science, Institute for the Modernization of Educational Content of Ukraine, National Academy of the Security Service of Ukraine, and the Research Institute of Informatics and Law of the National Academy of Legal Sciences of Ukraine, 2017, p. 106.

[63] Yu. V. Kovtun, "Zagrozi Informacijnomu Prostoru Derzhavi v Umovah Agresii Rosijs'koi Federacii [Threats to the State's Information Space Under the Conditions of the Aggression by the Russian Federation]," in *Current Problems Related to the Management of Information Security of the Government*, 8th Scientific and Applied Conference, Kyiv, Ukraine: Ukrainian Ministry of Education and Science, Institute for the Modernization of Educational Content of Ukraine, National Academy of the Security Service of Ukraine, and the Research Institute of Informatics and Law of the National Academy of Legal Sciences of Ukraine, 2017, p. 50.

[64] Kovtun, 2017, p. 50.

content that is allegedly created by Ukrainians. In an effort to bolster the legitimacy of this content, Russia also employs so-called experts who substantiate anti-Ukrainian narratives.[65] Relatedly, Levchenko documents the Kremlin's use of film, television, and other forums to engender positive perceptions of Russia in the eyes of Ukrainian audiences.[66]

Finally, the Ukrainian military-scientific literature frequently characterizes Russian information confrontation as complex and multifaceted. Zavodovs'ka, for example, characterizes Russian IPb as multidimensional, often targeting local, national, or international audiences that include decisionmakers, elites, the general public, or niche groups.[67] Zvarych offers a slightly more nuanced depiction of the targets of Russian IPb campaigns in Ukraine, which include the information infrastructure of the state; the morale and attitudes of armed services personnel and the broader public; decisionmaking systems within government, science, and industry; and the security sector.[68]

Russian Military Science Views of the Conflict in Ukraine

By contrast, the Russian military-scientific community has been notably tight-lipped on Russian actions in Ukraine. Those Russian military experts who address information campaigns targeting Ukraine do not openly acknowledge the Kremlin's role in these efforts. Rather, the consensus among Russian military-scientific thinkers—or at least their consensus as stated in publicly available sources—is that events in Ukraine are an example of Western hybrid warfare. Ukraine, Russian military experts suggest, has been the victim of Western information confrontation campaigns. Writing in 2014, for example, a group of senior Russian military officers cite both the Russo–Georgian War and the Ukraine conflict as evidence of recent Western information

---

[65] Kovtun, 2017, pp. 50–51.

[66] Levchenko, 2017, p. 13.

[67] Zavodovs'ka, 2015, p. 34.

[68] A. O. Zvarych, "The Experience of Ukrainian Counteraction to the Negative Information and Psychological Influence of the Russian Federation," *Zbirnyk Naukovykh Prats' Kharkivs'koho Natsional'noho Universytetu Povitryanykh Syl [Collection of Scientific Works of Kharkiv National University of the Air Force]*, Vol. 56, May 22, 2018, p. 9.

campaigns.[69] The prevailing narrative of "so-called 'Russian aggression'" in Georgia and Ukraine, they argue, is reflective of a "blatant falsification of events" rather than an accurate reflection of reality.[70] Based on their observations of the Georgia and Ukraine cases, they conclude that IPb "has acquired a more deceitful, hypocritical and aggressive character," a trend they describe as "especially evident in the current events in Ukraine."[71] Similarly, A. Bartosh, a correspondent for the Russian Academy of Military Sciences, claims that the United States and NATO are carrying out a "strategy of hybrid wars—phased multidimensional conflicts in . . . Ukraine."[72] The informational component of this strategy, according to Bartosh, has involved coopting "manipulated citizens" to stage a color revolution. In parallel, Western actors have combined special operations tactics with means of informational struggle.[73] This reflects a broader Russian narrative on Western-sponsored color revolutions as a tool for regime change and political influence.[74]

Some Russian military scholars focus more narrowly on the informational-psychological aspects of the West's alleged hybrid actions in Ukraine. Lieutenant Colonel V. V. Mikhailov and I. V. Puzenkin, for example, argue that American "informational-psychological operations in Ukraine serve as a striking example of information warfare in

---

[69] A. Ya. Fisun, V. K. Shamrey, A. Yu. Goncharenko, B. V. Ovchinnikov, and S. V. Chermianin, "Psihologija i Psihopatologija Informacionnyh Vojn [The Psychology and Psychopathology of Information Wars]," *Voenno-Medicinskij Zhurnal [Military-Medical Journal]*, No. 6, June 2014, pp. 4–12.

[70] Fisun, et al., 2014, pp. 4–5.

[71] Fisun, et al., 2014, pp. 4–5.

[72] A. Bartosh, "Gibridnye Ugrozy Zapada [Hybrid Threats from the West]," *Nezavisimoe Voennoe Obozrenie [Independent Military Review]*, No. 19, 2017, p. 1.

[73] Bartosh, 2017, p. 1.

[74] See, for instance, Roger McDermott, "Gerasimov Unveils Russia's 'Strategy of Limited Actions,'" *Real Clear Defense*, March 11, 2019; Andrei Soldatov and Michael Rochlitz, "The *Siloviki* in Russian Politics," in *The New Autocracy: Information, Politics, and Policy in Putin's Russia*, ed. Daniel Treisman, Washington, D.C.: The Brookings Institution, 2018.

recent decades."[75] Mikhailov and Puzenkin contend that the United States, with the cooperation of Ukrainian authorities in Kyiv, has attempted to malign the people of Donetsk and Luhansk by propagating false narratives about their involvement in the MH-17 incident. They maintain, moreover, that these campaigns have attempted to denigrate the inhabitants of Donetsk and Luhansk by labeling them as "terrorists" and "separatists."[76]

---

[75] I. V. Puzenkin and V. V. Mikhailov, "Rol' Informacionno-Psihologicheskih Sredstv v Obespechenii Oboronosposobnosti Gosudarstva [The Role of Informational-Psychological Means in Ensuring the Defense of the State]," *Voennaya Mysl' [Military Thought]*, Vol. 24, No. 3, 2015, pp. 1–6.

[76] Puzenkin and Mikhailov, 2015.

CHAPTER FIVE

Conclusions

Since the end of the Cold War, the Russian military-scientific literature has reflected the perception that Russia is behind the West in
its development of the theoretical underpinnings and operationaliza-
tion of information confrontation. The literature suggests that Russian military experts are concerned that Russia has consistently failed
to keep pace with the United States in the information domain. This
includes technological advancements; their application to military capabilities; the safeguarding of vulnerable systems, networks, and psyches
of armed personnel; and lastly, the development of IPb theory and
practice. Within the Russian military-scientific community, there have
been calls over the last three decades for the establishment of a coherent, unified doctrine of information confrontation and the standardization of key terms and their respective definitions. As of the writing
of this report, however, this standardization has not yet been achieved.

In keeping with this concern about the relative strengths of the
United States and Russia in competing via information confrontation,
the Russian military-scientific community has closely followed developments in American military doctrine and strategy pertaining to the
information domain, which has informed and shaped Russian perceptions of U.S. intentions and activities in the information space. The
profound influence of the Gulf War on the Russian military-scientific
community means that this has been particularly true since the early
1990s, when Russian experts witnessed what they perceived to be a
new generation of warfare unfolding between Iraq and U.S. and coalition forces, including the first demonstration of American information
confrontation. Russian military experts have also observed and drawn

lessons from other military operations including the Kosovo conflict, the Russo–Georgian War, and to a lesser extent, the Iraq War.

The importance of information confrontation in the eyes of Russian military-scientific thinkers emerged as a trend in our review of the Russian military-scientific literature. Russian military scholars often draw parallels between information weapons and WMD, a motif that illustrates the importance of IPb for this community. Both types of weapons are viewed as having the potential to affect massive changes within the international system. By fomenting dissent through an informational campaign, states can engineer the ousting of an unfavorable regime. In this conception, information weapons are like WMDs because they have the power to reshape regional and even global politics. Indeed, a number of articles within the Russian military-scientific literature characterize IPb as a means of achieving a "bloodless victory." In this sense, informational instruments allow states to prevail without ever directly engaging with the enemy. This notion of a "contactless war" appears to have been shaped by Russian observations of coalition operations during the Gulf War, which served as the first real-world demonstration of precision-guided weapons and the use of intelligence, surveillance, and reconnaissance to enable the conduct of psychological warfare.

Information confrontation plays a central role in shaping how Russia sees itself in the world. As a fundamental aspect of Russian foreign policy, it influences how Russia interacts with other international actors, including both allies and adversaries. Russia sees itself as being in a constant state of information confrontation with the West, as it tries to expand its own dominance and prevent its adversaries from gaining influence. Indeed, information confrontation has evolved from something that is primarily carried out during wartime, as a supplement to conventional military operations, into something that is carried out continuously, even in peacetime. It has gradually become a distinct and pervasive form of warfare. Information confrontation can be used to shape the operational environment in near-term conflicts, but it can also be used to ensure that the operational environment will be sufficiently malleable in future conflicts. According to Russian military experts, moreover, the informatization of society means that information confrontation is not going away. The growing role of information in all aspects of politics and society means that the more advanced a country is, the more vulnerable it will be to the effects of information confrontation.

Lastly, as the subject of one of Russia's most comprehensive information confrontation and hybrid warfare campaigns in the post–Cold War era, the Ukrainian experience offers valuable insights for Western audiences. In addition to the profound imprint that events since 2014 have left on Ukraine's military, political, economic, and social structures and processes, they have also shaped Ukrainian conceptions of information confrontation and influence campaigns. For instance, we observed a shift in tone between the pre- and post-Maidan eras in the SBU's literature on these concepts. The depiction of the 2008 Russo–Georgian conflict in the 2012 SBU textbook on information-psychological confrontation mirrors prevailing Russian narratives on the event. That is, the textbook frames Tbilisi as the aggressor and contends that the United States conducted informational campaigns painting Georgia as the victim with an eye toward shaping global public opinion in favor of Georgia.[1] By contrast, SBU writings published after the 2014 annexation of Crimea and ouster of the Yanukovych government do not appear to mirror the Russian government's narrative regarding its influence activities in Ukraine as closely. Unlike post-2014 Russian military-scientific writings, which do not explicitly acknowledge Russian information campaigns targeting Ukraine, SBU writings explicitly name Russia as the perpetrator—and in some cases, the sponsor—of informational campaigns targeting Ukraine.[2]

Recommendations

Based on the foregoing analysis of conceptions of information confrontation in the Russian military-scientific literature, we offer several recommendations. First, this literature provides a wealth of insight

---

[1]   Zharkov, 2012, pp. 153–155.

[2]   For instance see, Ivanov, 2017, pp. 38–40; Kovtun, 2017.

into the role of information in Russian military strategy and Russian perceptions of U.S. information operations. The U.S. intelligence community should study the Russian military-scientific literature to better understand Russian activities, intentions, and perceptions in the information domain. In addition, the U.S. military service intelligence offices and organizations should work more closely with the Open Source Enterprise to understand publicly available and unclassified Russian-language sources and encourage the use of the Russian military-scientific journals and Russian academic journals identified in this report.[3] The military services should also consider more closely monitoring Russian rhetoric regarding the distortion of historical facts, especially in Eastern Europe, as an emergent tool of information confrontation. Given the continuous nature of information confrontation as an element of Russian strategy, this analysis would assist in monitoring Russian influence operations and related activities. This literature can also be used to better understand how Russian perceptions of information confrontation and its role in modern warfare might influence resourcing, personnel, and procurement decisions within the Russian defense establishment.

Second, officials and observers have characterized Ukraine as a proving ground for Russian subversive activities, whether offensive cyberattacks or information confrontation campaigns, as well as for the demonstration and deployment of Russian weapons and military equipment and tactics.[4] As our research demonstrates, Ukraine also boasts a vibrant and prolific scholarly community that closely monitors developments pertaining to Russian subversive tactics, particularly as they unfold in the Ukrainian context. What is more, these scholars are actively involved in developing theoretical frameworks for information

---

[3] Although we acknowledge the limitations of open-source analysis in understanding Russian activities and concepts, we believe that the richness of the Russian military-scientific literature on IPb and related concepts provides valuable insights.

[4] "Experts Suspect Russia Is Using Ukraine as a Cyberwar Testing Ground," *NPR*, June 22, 2017; Defense Intelligence of the Ministry of Defense of Ukraine, "Russia's Military Aggression Against Ukraine: Ukraine—a Proving Ground for Modern Application of Prohibited Russian Weapons"; Geoffrey Cain, "Ukraine's War on Russian Disinformation Is a Lesson for America," *The New Republic*, March 29, 2019.

warfare and information confrontation—an issue that the U.S. defense establishment has long worked to (and continues to work to) untangle. Rather than continuing to focus its analysis on the observable operational characteristics of Russian information campaigns, the United States and the West would also benefit from opening their analytical aperture to incorporate the scholarly literature of Ukraine and other states that serve as litmus tests for Russian efforts. Establishing venues for continuous dialogue with prominent thinkers from Ukraine and other states that Russia perceives to be within its sphere of interest may also prove beneficial as a means of communicating and comparing observations and findings.

Further Research

The Russian military-scientific literature on information confrontation raises broader questions about the relationship between information confrontation, hybrid warfare, and other instruments of soft power. Additional research is needed to better understand the role of information confrontation in hybrid warfare and how it can be used as an instrument of soft power. In particular, observations and analyses of the Russian invasion of Ukraine in February 2022 may offer critical insight into the evolution of IPb as a component of Russian strategic thinking. Further research is also needed to understand how international governance mechanisms may be used to regulate information confrontation in the future.

Related Terms

Russian military scholars have identified inconsistencies in the terminology used to describe information confrontation and related concepts, such as information warfare, information weapons, information resources, information space, and information security. We define and provide additional background on a number of these related terms. We also provide a brief discussion of the use of these terms, as opposed to "information confrontation," in the Russian military-scientific literature.

Underlying Concepts

Informatization is a phenomenon that makes it possible to engage in information confrontation.[1] It serves as the foundation for the growing role of information activities and operations and their impact on modern society. Informatization entails "creating and developing a telecommunications infrastructure designed to unite geographically separated information resources."[2]

Discussions of **information aggression** in the military-scientific literature have primarily focused on the limitations of existing international norms and legal frameworks in countering it. Dylevskii,

---

[1] V. S. Shevtsov, "Informatsionnoye Protivoborstvo v Globaliziruyushemsia Mire: Aktual'nost', Differentsiatsiya Poniatiy, Ugrozy Politicheskoy Stabil'nosti [Information Confrontation in a Globalizing World: Relevance, Differentiation of Concepts, Threats to Political Stability]," *University Journal [Vestnik Universiteta]*, No. 5, 2015, pp. 206–211.

[2] Shevtsov, 2015, p. 206.

Komov, and Petrunin have argued that the definition of aggression in international agreements should be expanded to include aggression in the information space.[3] Even though the information space is transnational, information and communication technologies are owned by national actors and located within the physical borders of states. The disruption of information technologies therefore qualifies as an invasion of national sovereignty, territorial borders, and the political independence of another state.[4] Computer viruses like Stuxnet and distributed denial-of-service attacks in Estonia and Georgia have been cited as examples of the ability of information technologies to have transborder impacts, thereby making these technologies capable of carrying out acts of aggression.[5]

Bochkareva and Tsyganov have further hypothesized that entire socioeconomic systems, not just information technologies, can be the targets of information aggression.[6] In line with this hypothesis, Samokhvalov outlined the goals of information aggression, including a qualitative change in traditional cultural and spiritual life, the violation of the continuity of national ideals and values, and the dismantling of historical memory.[7] Information aggression can occur on different levels (e.g., at the level of the corporation, the region, the state, or even global society), but it exploits the same weaknesses within those entities (e.g., a lack of information, dissatisfaction with

---

[3] Dylevskii, Komov, and Petrunin, 2013, pp. 3–12. This source also references the term "cyber aggression," but in the context of its use in the National Security Strategy of the United States and the lack of definition for it in any official writings.

[4] Dylevskii, Komov, and Petrunin, 2013, p. 7.

[5] Dylevskii, Komov, and Petrunin, 2013, p. 7.

[6] Y. Bochkareva and V. Tsyganov, "Predposylki i Osobennosti Informatsionnykh Agressiy v Tsentralizovannykh i Liberal'nykh Sotsial'no-Ekonomicheskikh Sistemakh [Background and Characteristics of Information Aggression in Centralized and Liberal Social-Economic Systems]," *Informatsionniye Voyny [Information Wars Journal]*, Vol. 1, No. 29, 2014, pp. 75–81.

[7] V. I. Samokhvalov, "Spetsifika Sovremennoy Informatsionnoy Voyny: Sredstva i Tseli Porazheniya [Specifics of Modern Information War: Means and Purpose of Damage]," *Filosofiya i Obshestvo [Philosophy and Society]*, No. 3, July–September 2011, pp. 54–73.

a specific situation, or lack of education).[8] Bochkareva and Tsyganov conclude that states that are agile and have a good understanding of the theory and practice of information confrontation will succeed at information aggression.[9]

The proliferation of information weapons and the concept of information aggression have increased the vulnerability of a growing number of critical individual and state information resources. As a result of the phenomenon of informatization, developed countries have grown more dependent on the effectiveness of systems and the means of processing, disseminating, and using information, and this in turn has led to the concept of **state information potential**.[10] Molchanov defines this concept as the state's ability to secure the informatization of the society in which it is based.[11] **Military information potential**, on the other hand, is an element of the state's military potential. The quality of the military information potential determines the state's overall military potential and its ability to employ military power.[12]

The **single information space**, according to Russian Military Doctrine, is the sum of the information resources of the armed forces of the Russian Federation, as prescribed by the unified principles and rules of formation, formalization, storage, and dissemination.[13] The concept of network centric, or net-centric, warfare, as developed by the United States in the late 1990s, is evolving, and more recent scholarship has recommended that Russia update its definition of the single information space.

---

[8] Bochkareva and Tsyganov, 2014, p. 75.

[9] Bochkareva and Tsyganov, 2014, p. 75.

[10] Molchanov, 2008, pp. 2–9.

[11] Molchanov, 2008, p. 2.

[12] Molchanov, 2008, pp. 4–5.

[13] Chief of Staff of the Armed Forces of the Russian Federation, *Concept of Single Information Space of the Armed Forces of the Russian Federation*, December 16, 2004. Cited in V. V. Baranyuk and I. N. Ahmadishin, "Problemy Postroyeniya Yedinogo Informatsionnogo Prostranstva Vooruzhennykh Sil Rossiyskoy Federatsii i Vozmozhniye Puti ikh Resheniya [Problems with Building a Single Information Space of the Armed Forces of the Russian Federation and Possible Solutions]," *Voennaya Mysl' [Military Thought]*, No. 12, 2013, pp. 66–71.

Information War and Information Warfare

According to Russian military scholars, the term **information war** first appeared in Western literature in 1992. The Ministry of Defense of the Russian Federation defines information war as a "transparent and severe clash between states" that causes a "harmful impact on the information domain."[14] At the Army 2019 Conference, Minister of Defense Shoygu used the term when he noted the West's "aggressive information influence" on Russia.[15]

The official definition of information war, as set forth in the *Conceptual Views* document, has been incorporated by some academic scholars, but other academics argue that this definition is too broad.[16] Other scholars define information war more narrowly, as a struggle between opposing sides for superiority over the enemy in timeliness, assurance, completeness of information, speed, and quality of processing and dissemination.[17] Using this definition, Nuzhdin concludes that success in information war requires the following: "the maximum employment of resources, including non-traditional ones; discrediting and dehumanizing the competitor to the maximum extent possible; transforming the opposing side from the status of a 'competitor' to the status of 'enemy' or 'adversary'; and always staying ahead of the opponent, forcing them to explain themselves constantly, thus creating the perception of their guilt."[18] This definition, while narrow, does

---

[14] Ministry of Defense of the Russian Federation, "Informatsionnaya Voyna [Information War]," *Voyennyy Entsiklopedicheskiy Slovar' [Military Encyclopedic Dictionary]*, translated by Polina Kats-Kariyanakatte, Joe Cheravitch, Clint Reach, undated b.

[15] Ministry of Defense of the Russian Federation, 2019.

[16] See, for example, Dylevskii, Zapivakhin, Kovom, Korotkov, and Petrunin, 2014, pp. 3–12.

[17] Y. Nuzhdin, "Informatsionniye Voyny: Uroki Devianostykh [Information Wars: Lessons of the Nineties]," *Flag Rodiny [Flag of the Motherland]*, November 22, 2000.

[18] O. Nuzhdin, "Informatsionniye Voyni XV Veka: Preliudiya k Sovremennosi [Information Wars of the XV Century: Prelude to Modernity]," in *Informatsionniye Voyny kak Bor'ba Geopoliticheskikh Protivnikov, Tsivilizatsii i Razlichnykh Etosov [Information Wars as Struggle Between Geopolitical Opponents Civilizations and Ethos: Collection of Works of All-Russian Scientific Conference]*, Novosibirsk: Siberian State University Telecommunications and Information, April 26–27, 2018, pp. 517–527.

not consider damage to adversary information systems, processes and resources, and critical infrastructure to be a prerequisite for success in an information war. More generally, information war does not merely supplement traditional means of waging war; it is often at the forefront of achieving military aims.[19]

Some Russian military scientists treat **information warfare** as synonymous with information war. Suleimanov and Nazarova, for example, refer to "information warfare (war)."[20] Others use the term information warfare interchangeably with information confrontation. Dyleyskii, for example, identifies the "means of information confrontation (warfare) to include [the following]: means of technical reconnaissance, specially created or existing informational means, psychotronic means, means of special software effects, [and] means of information protection."[21] The academic literature has defined information warfare in terms of its goals—namely, to gain information superiority in the process of armed confrontation.[22] This definition is similar to that introduced by Nuzhdin in 2000 (and again by Slipchenko in 2002). More recently, Grudinin explained that information superiority ensures that "command and control over forces and equipment has more complete, precise, validated, and timely information about the operational environment than those of the enemy."[23]

Information warfare, like information confrontation, can be classified according to two subtypes: informational-technical and informational-

[19] V. Krasnoslobodtsev, Y. Kuzmin, A. Raskin, and I. Tarasov, "Informatsionnaya Bor'ba kak Osnovnoy Atribut Sovremennoy Voyny [Information Warfare as the Primary Attribute of Modern War]," *Informatsionniye Voyny [Information Wars Journal]*, Vol. 3, No. 39, 2016, p. 13.

[20] Sh. S. Suleimanova and E. A. Nazarova, *Informatsionniye Voyny: Istoriya i Sovremennost' [Information Wars: Past and Present]*, Moscow, 2017.

[21] Dylevskii, Zapivakhin, Kovom, Korotkov, and Krivchenko, 2016, pp. 3–11.

[22] M. A. Rodionov, 1998, pp. 67–70.

[23] I. Grudinin, "Effektivnost' Informatsionnogo Obespecheniya [Effectiveness of Information Operations]," *Armeiskii Sbornik [Army Digest]*, No. 11, November 2011, pp. 26–30.

psychological.[24] As with information confrontation, however, there are opposing views regarding whether information warfare includes both technical and nontechnical efforts. Krasnoslobodtsev and colleagues emphasize the technical nature of information warfare as distinguishing it from traditional warfare, because it encompasses the concept of "information-network warfare."[25] On the other hand, Orlansky argues that information warfare "cannot include any technical (power) aspect" and is conducted only for the purposes of exerting an informational-psychological influence."[26]

Tools and Techniques of Information Warfare

According to Limon and Krysanov, an effective system of information confrontation necessarily includes **concealment**. They adopt the following definition of concealment used in the Russian *Military Encyclopedic Dictionary*:

> A complex system of interrelated organizational, operational-tactical, engineering, and technical measures carried out in order to hide troops (fleet forces) and targets from the enemy and mislead him about the presence, location, composition, state, actions, and intentions of troops (fleet forces), and also command plans.[27]

---

[24] As noted in Chapter One, the informational-psychological subtype of IPb includes efforts to influence the enemy's population and military forces, including by "mislead[ing] the enemy, undermin[ing] its will to resist, produc[ing] panic in its ranks, and generat[ing] betrayal." The informational-technical subtype of IPb involves the physical manipulation of information networks and tools, including the "destruction of information, radio-electronic, [and] computer networks, and [gaining] unauthorized access to the information resources of the enemy." Trotsenko, 2016.

[25] Krasnoslobodtsev, et al., 2016, pp. 10–13.

[26] Orlansky, 2008, pp. 62–70.

[27] Limno and Krysanov, 2003, citing Ministry of Defense of the Russian Federation, "Maskirovka [Concealment]," *Voyennyy Entsiklopedicheskiy Slovar' [Military Encyclopedic Dictionary]*, undated d.

Concealment offers a "type of security for combat activities and the daily operations of troops (forces)," making it an important part of information operations in the context of military operations.[28]

Information campaigns are the means by which information effects are delivered. An information campaign is a planned flow of information, characterized by a prolonged period of time and intensity and having specific goals and tasks, which is then disseminated using different methods and channels to reach the masses and individuals.[29] Information campaigns are conducted to support the goals of political actors or to achieve political and informational goals in the process of winning an information battle.[30]

Rodionov further identified several key terms related to information warfare:[31]

- **Information operations** are a set of information battles, activities, and strikes coordinated in terms of purpose, objectives, place, and time that are conducted to gain or maintain information superiority over the enemy (offensive information operations) or reduce the enemy's information superiority (defensive information operations) in a given combat theater or strategic direction.

- **Information battles** can occur within the framework of information operations and aim to accomplish a vital operational task. An information battle is a set of information activities and strikes that are coordinated in terms of purpose, objectives, place, and time. An information operation can include several information battles occurring simultaneously or successively.

---

[28] Limno and Krysanov, 2003, pp. 70–74; Dylevskii, Zapivakhin, Kovom, Korotkov, and Petrunin, 2014, pp. 3–12; Ministry of Defense of the Russian Federation, undated d.

[29] Suleimanova and Nazarova, 2017, p. 87.

[30] Suleimanova and Nazarova, 2017, p. 59.

[31] Rodionov, 1998, pp. 68–70.

- **Information activities (actions)** are a set of activities that are carried out by forces and means involved in information warfare for a certain time in a given area (direction). Information activities can be classified according to their type (offensive and defensive), scale (strategic, operational-strategic, operational, operational-tactical, and tactical), and objects of influence (information and technical systems, the moral and psychological state of personnel, or some combination thereof).

Offensive information activities (actions) include information influence and information blockades, which are defined as follows:

- **Information influence** commences during peacetime, often long before the start of military operations. It supports effective information warfare, especially in such areas as intelligence analysis.[32] Actions associated with information influence go beyond the framework of information warfare, occurring within the sphere of "information confrontation of geopolitical actors."[33] This distinction suggests that information confrontation and information warfare are not synonymous, and that the latter is a narrower concept than the former. Information influence can consist of the manipulation of mass media, culture, and art.
- An **information blockade** is the use of forces and means to reduce the enemy's capability of obtaining and using the information necessary for the effective conduct of operations (combat activities). One of the main ways to achieve the goal of information blockade is through electronic blocking—a coordinated effect between means of electronic suppression and functional destruction of the technical elements of intelligence systems and channels for transmitting information. The goal of an information blockade cannot be fully achieved without special measures carried out by the country's military-political leadership. Within the framework of an information blockade, information strikes—of various types and scales—can also be carried out.

---

[32] Rodionov, 1998, p. 69.

[33] Rodionov, 1998, p. 69.

Defensive information activities (actions) are focused on information protection. They are defined as the coordinated use of forces and means involved in information warfare to ensure the stability of the operation of command and control of forces under conditions of information influence of the adversary.

- **Information strikes** can be carried out within the framework of information activities. They are understood as a short-term, powerful, and coordinated means of information influence over the most important element (or elements) of the adversary's command and control system to achieve decisive goals and gain information superiority (thereby reducing the adversary's information superiority). Information strikes can be classified according to their scale (strategic, operational-strategic, operational, operational-tactical, and tactical), type (radio-electronic, electronic fires strike, computer, special, and combined), and the degree to which they amass forces and means (selective, concentrated-massive, and massive).
- An **electronic strike** is the sudden, massive, complex impact of diverse forces and means of electronic suppression and the functional destruction of electronic objects of the adversary's control system, coordinated in time, depth, and objectives, in order to disrupt control for a certain period of time.
- An **electronic fires strike** is the massive, complex (radio-electronic and fires) impact of forces and means of electronic warfare, missile forces and artillery, aviation, and other forces and means allocated to the adversary's command and control systems. The goal of these strikes is to disrupt the adversary's C2 systems in certain areas for a certain period of time.
- A **computer (software) strike** is the sudden, massive, complex impact of attacking forces and means on the adversary's automated control system. It seeks to disrupt the adversary's control of that system.
- A **special strike** is a massive, complex moral and psychological impact of forces and means involved in information warfare on the personnel (primarily the personnel of C2 bodies) of adversary formations in order to disrupt or complicate the adversary's control of its personnel.

According to Rodionov, it is incorrect to consider information warfare outside the framework of specific strategic operations of the armed forces.[34] Grudinin also notes that information warfare must be conducted according to a unified plan during the preparation for and conduct of military operations.[35]

Information weapons have specific features that set them apart from conventional weapons. Bolotov defines an information weapon as a means of "destroying, distorting, [or] stealing information, extracting necessary information after dismantling defenses, limiting or restricting access to information by legitimate users, disrupting operation of technical means, [and] defeating communication networks [and] computer systems."[36] He also emphasized the unique features of information weapons, which set them apart from conventional weapons: the element of surprise, the scale of their application, their universal use, and their effectiveness.[37] Information weapons have been compared with WMD in that they should be subject to international legal norms to prevent their proliferation. Shehovtsov and Kuliashou have observed, for example, that the results of the application of information weapons are comparable to the use of WMD.[38] Dylevskii has also called for measures to counter the proliferation of information weapons.[39]

There is debate within the Russian military-scientific community about whether **propaganda** is a type of information weapon. While

---

[34] Rodionov, 1998, p. 67.

[35] Grudinin, 2011, p. 29.

[36] N. N. Bolotov, "Sushnost' i Soderzhaniye Poniatiya 'Voyna v Informatsionnoy Sfere' [The Essence and Content of the Concept of 'War in the Information Sphere']," *Vestnik Akademii Voennykh Nauk [Journal of the Academy of Military Sciences]*, Vol. 1, No. 58, 2017, pp. 22–28.

[37] Bolotov, 2017, p. 23; N. Shehovtsov and Y. Kuliashou, "Informatsionnoye Oruzhiye: Teoriya i Praktika v Informatsionnom Protivoborstve [Information Weapon: Theory and Application in Information Confrontation]," *Vestnik Akademii Voennykh Nauk [Journal of the Academy of Military Sciences]*, Vol. 1, No. 38, 2012, pp. 35–40.

[38] Shehovtsov and Kuliashou, 2012, pp. 35–40.

[39] Dylevskii, Zapivakhin, Kovom, Korotkov, and Krivchenko, 2016, pp. 5, 7.

some scholars have acknowledged that propaganda is used to exert a psychological influence on individuals, others have contended that it is not a weapon in the traditional sense because it does not possess the destructive characteristics of a traditional weapon.[40] Others have argued that propaganda is an example of a technology of information warfare that is comparable to information weapons.[41] Suleimanova and Nazarova define propaganda as "controlling mass consciousness by distorting information and providing one-sided, subjective, and often false ideas using means of mass information or other forms of mass effects" in order to influence public opinion and actions.[42] In their view, propaganda is "a 'network' of tricks [and] devices" such as advertisements and public relations in the hands of politicians.[43] Zorina, by contrast, classifies political advertisements, political agitation, and political public relations as a means of influencing public opinion that are distinct from propaganda.[44] Borkhsenius characterizes advertising agencies and public relations firms as instruments of soft power.[45]

Other Information-Based Types of Warfare

Network centric warfare is the conduct of combat operations to achieve information superiority to provide increased combat power of combined forces by creating a single information-communications

---

[40] Dylevskii, Zapivakhin, Kovom, Korotkov, and Petrunin, 2014, p. 5.

[41] Suleimanova and Nazarova, 2017, p. 51.

[42] Suleimanova and Nazarova, 2017, pp. 51–52.

[43] Suleimanova and Nazarova, 2017, p. 89.

[44] E. Zorina, "Propaganda kak Sovremenniy Instrument Vozdeystviya na Obshestvennoye Soznaniye [Propaganda as a Modern Instrument of Influence on Public Opinion]," *Informatsionniye Voyny [Information Wars Journal]*, Vol. 4, No. 36, 2015, pp. 89–93.

[45] Aleksandra Borkhsenius, "Information Warfare Operations. New Classification [Operatsii Informatsionnoy Voyny. Novaya Klassifikatsiya]," *Informatsionnyye Voyny [Information Wars]*, Vol. 39, 2016, p. 7.

network that connects sensors, decisionmakers, and warfighters.[46]
Network centric warfare consists of three subsystems: information,
intelligence (sensors), and combat.[47]

Russian strategic documents do not define the concept of **hybrid
warfare**, and at least until 2015, Russian officials rejected the concept of hybrid warfare as a descriptor of Russian activities.[48] Matvienko
traces the origins of hybrid or "multimodal" wars to what he characterizes as U.S. and NATO aggression in Yugoslavia.[49] In 2016, Gerasimov observed that "hybrid war requires high-tech weapons and scientific justification" to support the use of minimal armed forces against
the enemy.[50] He concluded that traditional and hybrid methods had
become common features of any armed conflict.[51]

During hybrid conflicts, according to Gorshechnikov, a confrontation begins long before the commencement of armed hostilities.[52]
At any moment, and in response to any "insignificant" situation, he
writes, an Arab Spring or color revolution can arise.[53] Color revolutions, indirect action, and activities associated with soft power typically

---

[46] Baranyuk and Ahmadishin, 2013, p. 67.

[47] V. Zinoviev, A. Koldunov, and N. Gruzdew, "Perspektivy Primeneniya Informatsionnykh
Setey v Voyennom Dele [Possible Uses of Information Networks in Military Activities],"
Informatsionniye Voyny [Information Wars Journal], Vol. 1, No. 33, 2015, pp. 37–40.

[48] Timothy L. Thomas, *Russia Military Strategy: Impacting 21st Century Reform and Geopolitics*, Fort Leavenworth, Kan.: Foreign Military Studies Office (FMSO), 2015, p. 83.

[49] Yu. Matvienko, "'Tsvetniye' Revoliutsii kak Nevoyenniy Sposob Dostizheniya Politicheskikh
Tseley v Gibridnoy Voyne: Sushnost', Soderzhaniye, Vozmozhniye Mery Zashity i Protivodeystviya ['Color' Revolutions as Non-Military Means to Achieve Political Goals in
'Hybrid' War: Nature, Content, Possible Protection Measures and Countermeasures]," *Informatsionniye Voyny [Information Wars Journal]*, Vol. 4, No. 40, 2016, pp. 11–19.

[50] Valery Gerasimov, "Po Opytu Sirii [Syrian Experience]," *Voenno-Promyshlennii Kur'er
Online [Military-Industrial Courier Online]*, March 7, 2016.

[51] Gerasimov, 2016.

[52] O. M. Gorshechnikov, A. I. Malyshev, and Yu. F. Pivovarov, "Problemy Tipologii Sovre-
mennykh Voyn i Vooruzhennykh Konfliktov [Problems of Typology of Modern Wars and
Armed Conflicts]," *Vestnik Akademii Voyennykh Nauk [Journal of the Academy of Military
Sciences]*, Vol. 1, No. 58, 2017, p. 52.

[53] Gorshechnikov, Malyshev, and Pivovarov, 2017, p. 52.

appear in the context of what is termed **geopolitical confrontation**.[54] According to Yarkov, geopolitical confrontation can take five different forms: war using means of armed warfare; war using nonmilitary measures; armed conflict; conflict with the use of nonmilitary measures; and natural competition. He stipulates that nonmilitary measures can include economic, political, political-diplomatic, social, legal, and informational, including informational-technical and informational-psychological, activities.[55]

According to Karjakin, indirect action and soft power, which are also called **organization weapons** by some Russian military scholars, are the "most effective methods for the conduct of geopolitical warfare [or] struggle" and are frequently used to weaken an adversary.[56] More recent scholarship, however, draws a distinction between indirect action and soft power. For example, Morozov has observed that the United States uses soft power to maintain its geopolitical influence in other countries and promote pro-American views.[57] This use of soft power is not intended to be an adjunct to the political, economic, diplomatic, and military elements of U.S. foreign policy, but rather represents an independent element of the foreign policy aims. The United States employs a strategy of indirect action when it topples unfavorable regimes in countries around the world.[58] Indirect action and soft power

---

[54] V. Karjakin, "Strategii Nepriamykh Deystviy, 'Miagkoy Sily' i Tekhnologii 'Upravliay-emogo Haosa' kak Instrumenty Pereformatirovaniya Politicheskikh Prostranstv [Strategies of Indirect Action, 'Soft Power' and Technologies of 'Controlled Chaos' as Instruments of Reformatting of Political Spaces]," *Informatsionniye Voyny [Information Wars Journal]*, Vol. 3, No. 31, 2014, pp. 29–38.

[55] S. A. Yarkov, "Nevoyennyye Sredstva i Nevoyennyye Mery Neytralizatsii Voyennykh Opasnostey: Sushnostnoye Razlichiye i Predmetnaya Khrakteristika Poniatiy [Non-Military Means and Non-Military Measures to Neutralize Military Dangers: Essential Difference and Objective Characteristics of the Concepts]," *Natsional'naya Bezopasnost' [National Security]*, No. 3, 2017, p. 3.

[56] V. Karjakin, 2014, pp. 29–38.

[57] Yu. Morozov, "Primeneniye SShA 'Miagkosilovogo Arsenala' v Sovremennom Mire [Application of 'Soft Power Arsenal' by the USA in the Modern World]," *Informatsionniye Voyny [Information Wars Journal]*, Vol. 1, No. 41, 2017, pp. 16–24.

[58] Morozov, 2017, p. 18.

are employed in color revolutions, which can be classified as military-political confrontations in which the ultimate goal is the defeat of the adversary by an aggressor.[59] Information warfare has been employed in indirect action strategies in order to control the adversary state by destabilizing its political system.[60]

Alternative terms for informational-psychological confrontation include **conscientious war** and **behavioral war**, although these terms are less used in the Russian military-scientific literature. Dolgopolov, for example, associates the concept of conscientious war with "alteration [or] hacking of the consciousness of the opposite side."[61] Matvienko references the notion of "behavioral war," explaining that the "technology of behavioral control is the basis of the so-called conscientious war or war to defeat consciousness."[62] He characterizes behavioral war as a "means of interstate confrontation of tomorrow . . . based on technologies for manipulating behavioral algorithms, habits, [and] stereotypes of activity, embedded in us by society in the broadest sense of the word."[63]

Larina and Ovchinsky describe the development of new types of behavioral weapons that stem from Big Data technologies and intelligent computing, as well as the latest developments in social and behav-

---

[59] A. V. Dolgopolov, "Sovremennoye Ponimaniye Sushchnosti i Soderzhaniya Voyny [Current Understanding of the Essence and Content of War]," *Vestnik Akademii Voennykh Nauk [Journal of Academy of Military Sciences]*, Vol. 58, 2017, p. 45.

[60] A. S. Semchenkov, "Protivodeystviye Sovremennym Ugrozam Politicheskoy Stabil'nosti v Sisteme Obespecheniya Natsional'noy Bezopasnosti Rossii [Responding to Modern Threats to Political Stability in Russia's National Security System]," dissertation, Moscow Federal University of M. V. Lomonosov, 2012, p. 40. Cited in I. Il'in and S. Bilyuga, "Destabilizatsiya Sotsial'no-Politicheskikh Sistem: Osnovniye Podkhody k Poniatiynomu Apparatu [Destabilization of Socio-Political Systems: Basic Approaches to Conceptual Apparatus]," *Informatsionniye Voyny [Information Wars Journal]*, Vol. 4, No. 44, 2017, pp. 31–34.

[61] Dolgopolov, 2017, p. 45.

[62] Yu. A. Matvienko, "Nevoyennyye Ugrozy kak Sostavnaya Chast' Sovremennogo Mezhgosudarstvennogo Protivoborstva [Non-Military Threats as a Part of Contemporary Interstate Confrontation]," *Vestnik Akademii Voennykh Nauk [Journal of the Academy of Military Science]*, Vol. 1, No. 58, 2017, p. 37.

[63] Matvienko, 2017, p. 37.

ioral psychology.[64] Kugusheva similarly describes the basic elements of a new type of behavioral weapon called "the Nudge."[65] Nudging, which involves the use of habits and stereotypes to nudge a person or group of people toward making certain decisions and taking specific, desirable actions as a result of those decisions, represents a new tool for programming and controlling human behavior.[66] Kugusheva has also claimed that the United States is hiding a universal weapon called the "Supernudge" that it plans to use to achieve world domination.[67]

Information Confrontation Beyond the State

Information extremism is a "hard, radical position," according to Matveeva and Nosova, and "propaganda of extreme political measures."[68] They characterize the primary elements of information extremism as radical activities, antisocialism, immorality, the distortion of political-legal thinking, and illegitimacy.[69] These characteristics are suggestive of what information extremism *is* and what information extremism *does*. Information extremism has been further defined as a destructive phenomenon of modern society.[70] Information extremism can help

---

[64] E. Larina and V. Ovchinsky, "Novaya Voyennaya Strategiya SShA i Povedencheskiye Voyny [New U.S. Military Strategy and Behavioral Wars]," *Informatsionniye Voyny [Information Wars Journal]*, Vol. 3, No. 35, 2015, pp. 27–33.

[65] A. Kugusheva, "Ot Informatsionnykh Voyn k Povedencheskim [From Information Wars to Behavioral Ones]," *Informatsionniye Voyny [Information Wars Journal]*, Vol. 1, No. 37, 2016, pp. 11–22.

[66] Kugusheva, 2016, p. 19.

[67] Kugusheva, 2016, p. 21.

[68] E. Y. Matveeva and I .V. Nosova, "Informatsionniy Extremism: Sushnost' i Proyavleniya [Information Extremism: Characteristics and Manifestations]," in *Informatsionniye Voyny kak Bor'ba Geopoliticheskikh Protivnikov, Tsivilizatsii i Razlichnykh Etosov [Information Wars as Struggle Between Geopolitical Opponents Civilizations and Ethos: Collection of Works of All-Russian Scientific Conference]*, Novosibirsk: Siberian State University Telecommunications and Information, April 26–27, 2018, pp. 434–443.

[69] Matveeva and Nosova, 2018, p. 437.

[70] Matveeva and Nosova, 2018, p. 435.

achieve the goals of information aggression by carrying out information strikes in four different areas:

1. information-semantic, to disorient the individual;
2. information-emotional, to destroy the individual's ability to process events;
3. information-moral, to destroy the individual's ability to perceive the differences between good and evil; and
4. information-historic, to obliterate historical memory.[71]

Information terrorism refers to a narrower concept that is related to cyberterrorism and the internet activities of terrorist organizations. Ibragimov, for example, has observed that the use of internet-based communications is becoming one of the key instruments of terrorist organizations in supporting their activities through individual, financial, and information dealings across the internet.[72] Borkhsenius has noted that religious terrorist organizations have successfully conducted influence operations as part of an "informational-psychological war against the entire world."[73]

---

[71] Matveeva and Nosova, 2018, p. 435.

[72] L. Ibragimov, "Internet-Terrorism kak Fenomen Sovremennykh Politicheskikh Kommunikatsiy [Internet-Terrorism as a Phenomenon of Modern Political Communication]," *Informatsionniye Voyny [Information Wars Journal]*, Vol. 2, No. 38, 2016, pp. 71–75.

[73] Borkhsenius, 2016, p. 8.

References

Al'tshuller, R. E., and A. G. Tartakovskiy, *Listovki Otechestvennoy Voyny 1812 Goda [Leaflets of the Patriotic War of 1812]*, U.S.S.R. Academy of Sciences, 1962.

Adamsky, Dmitry, "Continuity in Russian Strategic Culture: A Case Study of Moscow's Syria Campaign," *Security Insights*, Marshall Center, No. 048, February 2020.

Antonovich, P. I., "Izmeneniye Vzgliadov na Informatsionnoye Protivoborstvo na Sovremennom Etape [Changing Views of Information Confrontation in the Modern Era]," *Vestnik Akademii Voennykh Nauk [Journal of the Academy of Military Sciences]*, Vol. 34, No. 1, 2011.

Astashov, A. B., *Propaganda na Russkom Fronte v Gody Pervoy Mirovoy Voyny [Propaganda on the Russian Front During the First World War]*, Spetskniga, 2012.

Baranyuk, V. V., and I. N. Ahmadishin, "Problemy Postroyeniya Yedinogo Informatsionnogo Prostranstva Vooruzhennykh Sil Rossiyskoy Federatsii i Vozmozhniye Puti ikh Resheniya [Problems with Building a Single Information Space of the Armed Forces of the Russian Federation and Possible Solutions]," *Voennaya Mysl' [Military Thought]*, No. 12, 2013.

Bartosh, A., "Gibridnye Ugrozy Zapada [Hybrid Threats from the West]," *Nezavisimoe Voennoe Obozrenie [Independent Military Review]*, No. 19, 2017.

Bikkenin, R., "Informatsionnoye Protivoborstvo v Voyennoy Sfere [Information Conflict in the Military Sphere: Basic Elements and Concepts]," *Morskoi Sbornik [Naval Digest]*, No. 10, 2003.

Bocharnikov, V. P., and S. V. Svyeshnikov, "Pohlyady na Kharakter Suchasnykh Voyennykh Konfliktiv [Views on the Character of Modern Military Conflicts]," *Nauka i Oborona [Science and Defense]*, No. 1, 2017.

Bochkareva Y., and V. Tsyganov, "Predposylki i Osobennosti Informatsionnykh Agressiy v Tsentralizovannykh i Liberal'nykh Sotsial'no-Ekonomicheskikh Sistemakh [Background and Characteristics of Information Aggression in Centralized and Liberal Social-Economic Systems]," *Informatsionniye Voyny [Information Wars Journal]*, Vol. 1, No. 29, 2014.

Bogdanov, S. A., "Veroyatnyj Oblik Vooruzhennoj Bor'by Budushego [Likely Features of the Armed Struggle of the Future]," *Voennaya Mysl' [Military Thought]*, No. 12, 2003.

Bolotov, N. N., "Sushnost' i Soderzhaniye Poniatiya 'Voyna v Informatsionnoy Sfere' [The Essence and Content of the Concept of 'War in the Information Sphere']," *Vestnik Akademii Voennykh Nauk [Journal of the Academy of Military Sciences]*, Vol. 1, No. 58, 2017.

Borkhsenius, Aleksandra, "Information Warfare Operations. New Classification [Operatsii Informatsionnoy Voyny. Novaya Klassifikatsiya]," *Informatsionnyye Voyny [Information Wars]*, Vol. 39, 2016.

Burtsev, M. I., *Prozrenie [Epiphany]*, Voenizdat, 1981.

Butuzov, Vitalij Mykolajovych, and Kateryna Viktorivna Titunina, "Suchasni Zahrozy: Komp'yuternyj Teroryzm [Current Threats: Computer Terrorism]," *Borot'ba z Orhanizovanoyu Zlochynnistyu i Korupciyeyu [Fight Against Organized Crime and Corruption]*, No. 17, 2007.

Cain, Geoffrey, "Ukraine's War on Russian Disinformation Is a Lesson for America," *The New Republic*, March 29, 2019.

Chekinov, S. G., and S. A. Bogdanov, "Strategicheskoe Sderzhanie i Nacional'naa Bezopasnost' Rossii na Sovremennom Etape [Strategic Deterrence and National Security, Russia in the Modern Period]," *Voennaya Mysl' [Military Thought]*, No. 3, 2012.

Chernichenko, Stanislav, "Ideologicheskaya Agressiya kak Primeneniye Sily v Mezhdunarodnom Prave [Ideological Aggression as the Use of Force in International Law]," *Ievraziyskiy Iuridicheskiy Zhurnal [Eurasian Law Journal]*, Vol. 128, No. 1, 2019.

Chernykh, G. S., "Ob Uchastii Vojsk RHB Zashity v Informacionnom Protivoborstve [On the Participation of RCB Protection Troops in Information Confrontation]," *Voennaya Mysl' [Military Thought]*, No. 6, 2002.

Chief of Staff of the Armed Forces of the Russian Federation, *Concept of Single Information Space of the Armed Forces of the Russian Federation*, December 16, 2004.

Chizhevskii, Y. A., "Osnovniye Tendentsii Transformatsii Prirody i Kharaktera Sovremennykh Voyenno-Politicheskikh Konfliktov [Main Trends in the Transformation of the Nature and Character of Contemporary Military-Political Conflicts]," *Voennaya Mysl' [Military Thought]*, July 11, 2020.

Darczewska, Jolanta, "Russia's Armed Forces on the Information War Front: Strategic Documents," OSW (Center for Eastern Studies), No. 57, June 2016.

Decree of the President of the Russian Federation, "On the Approval of the Military Doctrine of the Russian Federation," N 706, April 21, 2000.

Defense Intelligence of the Ministry of Defense of Ukraine, "Russia's Military Aggression Against Ukraine: Ukraine—a Proving Ground for Modern Application of Prohibited Russian Weapons," undated.

"Deistviia Soedninenii Chastei i Podrazdelenii SV Pri Provedenii Spetsial'noi Operatsii po Razoruzheniiu NVF v 1994–96 gg na Territorii Chechenskoi Respubliki Spisok Sokrashchenii i Abbreviatur [Actions of Divisions and Subdivisions of the Army During a Special Operation to Disarm Illegal Armed Groups in 1994–96 on the Territory of the Chechen Republic. List of Abbreviations and Acronyms]," *Doklad Byvshevo Nachal'nika Shtaba SKVO General-Leitenant V. Potapova [Report of the Former Chief of the North Caucasus Military District, Lieutenant-General V. Potapov]*, Vestnik PVO, April 11, 2005. As of June 7, 2021: http://pvo.guns.ru/book/chechnya_pvo.htm

Department of Defense Directive 4600.4, "Electronic Warfare (EW) and Command, Control, and Communications Countermeasures (C3CM)," Washington, D.C.: Department of Defense, August 27, 1979.

Derbin, E. A., "Metodicheskye Aspekti Sushnosti Sovremennykh Voin [Methodological Aspects of the Analysis of the Essence of Modern Wars]," *Vestnik Akademii Voyennykh Nauk [Journal of the Academy of Military Sciences]*, Vol. 1, No. 58, 2017.

"Disinformation Report on Foreign Interference in the 2016 Election," Yonder website, December 17, 2018. As of January 25, 2021: https://www.yonder-ai.com/resources/the-disinformation-report/

Dolgopolov, A. V., "Sovremennoye Ponimaniye Sushchnosti i Soderzhaniya Voyny [Current Understanding of the Essence and Content of War]," *Vestnik Akademii Voennykh Nauk [Journal of Academy of Military Sciences]*, Vol. 58, 2017.

Doroshenko, Margarita, and Daniil Kazarin, "Mezhdunarodno-Pravovoye Regulirovaniye Deyatel'nosti SMI v Usloviyakh Informatsionnykh Voyn [International Legal Regulation of Media Activities in the Context of Information Wars]," *Molodoy Uchenyy [Young Scientist]*, No. 141, 2017.

Dvornikov, Aleksandr, "Shtaby Dlya Novykh Voyn [Headquarters for the New Wars]," *Voenno-Promyshlennii Kur'er [Military-Industrial Courier]*, No. 28, 2018.

Dylevskii, I. N., S. A. Komov, and A. N. Petrunin, "Ob Informatsionnykh Aspektakh Mezhdunarodno-Pravovogo Poniatiya 'Agressiya' [On Information Aspects of the International Legal Definition of 'Aggression']," *Voennaya Mysl' [Military Thought]*, No. 10, 2013.

Dylevskii, I. N., V. O. Zapivakhin, S. A. Kovom, S. V. Korotkov, and A. A. Krivchenko, "O Dialekte Sderzhivaniya i Predotvrasheniya Voyennykh Konfliktov v Informatsionnuyu Eru [On the Dialect of Deterrence and Prevention of Military Conflicts in the Information Era]," *Voennaya Mysl' [Military Thought]*, No. 7, 2016.

Dylevskii, I. N., V. O. Zapivakhin, S. A. Komov, S. V. Korotkov, and A. N. Petrunin, "Mezhdunarodniy Rezhim Nerasprostraneniya Informatsionnogo Oruzhiya: Utopiaya ili Real'nost'? [International Regime for Nonproliferation of Information Weapons: Utopia or Reality?]," *Voennaya Mysl' [Military Thought]*, Vol. 10, 2014.

Dziuba, M., Ya. M. Zharkov, I. O. Olkhovoi, and M. I. Onyshchuck, *Narys Teorii i Praktyky Informatsiino—Psykholohichnykh Operatsii [Essay on the Theory and Practice of Information and Psychological Operations]*, Kyiv, Ukraine: VITI NTUU "KPI," 2006.

Efanov, A. A., "Izvechnoye Rossiysko-Amerikanskoye Protivostoyaniye, ili O Chetyrekh Etapakh Odnoy Informatsionnoy Voyny v Noveyshey Istorii [The Eternal Russian-American Confrontation, or About Four Stages of One Information War in Modern History]," in *Informatsionniye Voyny kak Bor'ba Geopoliticheskikh Protivnikov, Tsivilizatsii i Razlichnykh Etosov [Information Wars as Struggle Between Geopolitical Opponents Civilizations and Ethos: Collection of Works of All-Russian Scientific Conference]*, Novosibirsk: Siberian State University Telecommunications and Information, April 26–27, 2018, pp. 217–224.

Elkin, A. P., and A. I. Starinkov, "K Voprosu ob Intellektual'nykh Sistemakh Upravleniya [On the Question of Intelligent Control Systems]," *Voennaya Mysl' [Military Thought]*, No. 1, 1992.

"Experts Suspect Russia Is Using Ukraine as a Cyberwar Testing Ground," *NPR*, June 22, 2017.

Facon, Isabelle, *Reforming Ukrainian Defense: No Shortage of Challenges*, Paris: IFRI, No. 101, 2017.

Fadeev, A. S., "Voyenniye Konflicty Sovremennosti, Perspektivy Razvitiya Sposobov ikh Vedeniya, Priamiye i Nepriamiye Deistviya v Kofliktakh XXI Veka [Military Conflicts of Modernity, Prospects for the Development of the Ways of Their Management. Direct and Indirect Actions in Armed Conflicts of the 21st Century]," *Voennaya Mysl' [Military Thought]*, November 7, 2019. As of January 6, 2021:
https://vm.ric.mil.ru/Stati/item/222832/

"Fake News or the Truth? Russia Cracks Down on Virus Postings," *Associated Press*, April 1, 2020.

Filipenko, Kira, "Rosiiskiie Sily Informatsionno-Psikhologicheskikh Operatsii v Belorussii [Russian Forces of Information-Psychological Operations in Belarus]," InoSmi.ru, September 13, 2017. As of April 6, 2021:
https://inosmi.ru/military/20170913/240268097.html

Fisun, A. Ya., V. K. Shamrey, A. Yu. Goncharenko, B. V. Ovchinnikov, and S. V. Chermianin, "Psihologija i Psihopatologija Informacionnyh Vojn [The Psychology and Psychopathology of Information Wars]," *Voenno-Medicinskij Zhurnal [Military-Medical Journal]*, No. 6, June 2014.

Fitzgerald, Mary C., "The Soviet Image of Future War: Through the Prism of the Persian Gulf," *Comparative Strategy*, Vol. 10, No. 4, 1991.

Fraher, John, and Ilya Arkhipov, "Putin Says Patriotic Hackers Could Be Fighting for Russia," Bloomberg, June 1, 2017. As of January 25, 2021:
https://www.bloomberg.com/news/articles/2017-06-01/putin-says-patriotic-hackers -could-be-fighting-for-russia-j3eb8n22

Gallagher, Stuart, "Mission Ukraine: U.S. Army Leads Multinational Training Group to Counter Russian Threat," *Association of the United States Army*, May 19, 2020.

Gareev, M. A., and N. I. Turko, "Voyna: Sovremennoye Tolkovaniye Teorii i Realii Praktiki [War: Modern Interpretation of Theory and Realities of Practice]," *Vestnik Akademii Voyennykh Nauk [Journal of the Academy of Military Sciences]*, Vol. 1, No. 58, 2017.

Gerasimov, V., "Cennost' Nauki v Predvidenii [The Value of Science in Foresight]," *Voenno-Promyshlennii Kur'er [Military-Industrial Courier]*, 2013.

Gerasimov, Valery, "Po Opytu Sirii [Syrian Experience]," *Voenno-Promyshlennii Kur'er Online [Military-Industrial Courier Online]*, March 7, 2016.

———, "Vectory Razvitiya Voyennoy Strategii [Military Strategy Development Vectors]," *Krasnaya Zvezda [Red Star]*, March 4, 2019.

Gerasymchuk, Sergiy, *Research Report on the Second Investigation Level: Ukrainian Case*, Frankfurt, Germany: Peace Research Institute Frankfurt, 2008.

Giles, Keir, "'Information Troops'—a Russian Cyber Command?" In *3rd International Conference on Cyber Conflict*, eds. C. Czosseck, E. Tyugu, and T. Wingfield, Tallinn, Estonia: CCD COE Publications, 2011.

Gorbunov, Evgenii, *Stalin i GRU: 1918–1941 gody [Stalin and the Main Intelligence Directorate: 1918–1941]*, Rodina, 2018.

Gorshechnikov, O. M., A. I. Malyshev, and Yu. F. Pivovarov, "Problemy Tipologii Sovremennykh Voyn i Vooruzhennykh Konfliktov [Problems of Typology of Modern Wars and Armed Conflicts]," *Vestnik Akademii Voyennykh Nauk [Journal of the Academy of Military Sciences]*, Vol. 1, No. 58, 2017.

Grinyaev, S., "Tochka Zreniya: Informatsionnoye Prevoskhodstvo Vmesto 'Yadernoy Dubinki' [Opinion: Information Superiority Versus the 'Nuclear Stick']," *Armeiskii Sbornik [Army Digest]*, No. 5, 2002.

Grisé, Michelle, Yuliya Shokh, Khrystyna Holynska, and Alyssa Demus, *Russian and Ukrainian Perspectives on the Concept of Information Confrontation: Translations, 2002–2020*, Santa Monica, Calif.: RAND Corporation, RR-A198-7, 2021.

Grudinin, I., "Effektivnost' Informatsionnogo Obespecheniya [Effectiveness of Information Operations]," *Armeiskii Sbornik [Army Digest]*, No. 11, November 2011.

Gryzlov, B. M., and A. B. Pertsev, "Informatsionnoye Protivoborstvo: Istoriya i Sovremennost [Information Confrontation: History and Modernity]," *Vestnik Akademii Voennykh Nauk [Journal of the Academy of Military Sciences]*, No. 2, 2015.

Gurevich, P. S., *Propaganda v Ideologicheskoy Bor'be [Propaganda in Ideological Confrontation]*, Vyshaya shkola, 1987.

Gusev, Yurii Petrovich, "Spetsial'naia Propaganda v Moei Zhizni [Special Propaganda in My Life]," undated. As of January 28, 2020: http://textarchive.ru/c-2468205-pall.html

Hai-Nyzhnyk, Pavlo, ed. *Ahresiia Rosii Proty Ukrainy: Istorychni Peredumovy ta Suchasni Vyklyky [Russia's Aggression Against Ukraine: Historical Background and Current Challenges]*, Kyiv, Ukraine: MP Lesya, 2016.

Helmus, Todd C., Elizabeth Bodine-Baron, Andrew Radin, Madeline Magnuson, Joshua Mendelsohn, William Marcellino, Andriy Bega, and Zev Winkelman, *Russian Social Media Influence: Understanding Russian Propaganda in Eastern Europe*, Santa Monica, Calif.: RAND Corporation, RR-2237-OSD, 2018. As of May 10, 2021: https://www.rand.org/pubs/research_reports/RR2237.html

Higgins, Andrew, "As Putin Era Begins to Wane, Russia Unleashes a Sweeping Crackdown," *New York Times*, October 24, 2019.

Hodgson, Quentin E., Logan Ma, Krystyna Marcinek, and Karen Schwindt, *Fighting Shadows in the Dark: Understanding and Countering Coercion in Cyberspace*, Santa Monica, Calif.: RAND Corporation, RR-2961-OSD, 2019. As of May 10, 2021: https://www.rand.org/pubs/research_reports/RR2961.html

Horbulin, Volodymyr, *Svitova Hibrydna Viina: Ukrainskyi Front [The World Hybrid War: Ukrainian Forefront]*, Kyiv, Ukraine: National Institute for Strategic Studies, 2017.

Hutcherson, Norman B., *Command and Control Warfare: Putting Another Tool in the War-Fighter's Data Base*, Maxwell Air Force Base, Ala: Air University Press, September 1994.

Iasiello, Emilio J., "Russia's Improved Information Operations: From Georgia to Crimea," *Parameters*, Vol. 2, No. 47, 2017.

Ibragimov, L., "Internet-Terrorism kak Fenomen Sovremennykh Politicheskikh Kommunikatsiy [Internet-Terrorism as a Phenomenon of Modern Political Communication]," *Informatsionniye Voyny [Information Wars Journal]*, Vol. 2, No. 38, 2016.

Il'in, I., and S. Bilyuga, "Destabilizatsiya Sotsial'no-Politicheskikh Sistem: Osnovniye Podkhody k Poniatiynomu Apparatu [Destabilization of Socio-Political Systems: Basic Approaches to Conceptual Apparatus]," *Informatsionniye Voyny [Information Wars Journal]*, Vol. 4, No. 44, 2017.

"Information Troops Created in Russia [V Rossii Sozdani Voiska Infromacionnih Operacii]," *RIA Novosti*, February 22, 2017. As of May 15, 2021: https://ria.ru/20170222/1488596879.html?in=t

"Istochnik v Minoborony: V Vooruzhennykh Silakh RF Sozdany Voiska Informatsionnykh Operatsiy [Source at the Ministry of Defense: The Armed Forces of the Russian Federation have Created Information Operations Forces]," *TASS*, May 12, 2014.

Ivanov, O. Yu., "Rosijs'ko-Ukrains'ke Informacijne Protiborstvo z «Krims'kogo Pitannia»: Genezis Ta Suchasnij Stan [Russian-Ukrainian Confrontation and the 'Crimea Question': Genesis and Current Status]," in *Current Problems Related to the Management of Information Security of the Government*, 8th Scientific and Applied Conference, Kyiv, Ukraine: Ukrainian Ministry of Education and Science, Institute for the Modernization of Educational Content of Ukraine, National Academy of the Security Service of Ukraine, and the Research Institute of Informatics and Law of the National Academy of Legal Sciences of Ukraine, 2017, pp. 38–40.

Ivanov, Vladimir, "Samye Informirovannye Ludi v GRU [The Most Informed People in the GRU]," *Nezavisimaya Gazeta [Independent Gazette]*, October 12, 2012.

Joint Chiefs of Staff, *Joint Publication 13-3*, "Information Operations," November 20, 2014.

Karjakin, V., "Strategii Nepriamykh Deystviy, 'Miagkoy Sily' i Tekhnologii 'Upravliayemogo Haosa' kak Instrumenty Pereformatirovaniya Politicheskikh Prostranstv [Strategies of Indirect Action, 'Soft Power' and Technologies of 'Controlled Chaos' as Instruments of Reformatting of Political Spaces]," *Informatsionniye Voyny [Information Wars Journal]*, Vol. 3, No. 31, 2014.

Karpuchin, V., "Informatsionnaya Bezopasnost' Voysk [Information Security of Troops]," *Boevaya Vakhta [Battle Watch]*, No. 23, March 28, 2007.

Kaverin, Mikhail, "Liderstvo Rossiyskoy Federatsii v Sisteme Mezhdunarodnykh Institutov [Leadership of the Russian Federation in the System of International Institutions]," *Informatsionnyye Voyny [Information Wars]*, Vol. 46, No. 2, 2018.

Kenispaev, Zh. K., and N. S. Serova, "Civilizatsionnye Voiny: Antichnost [Civilization Wars: Antiquity]," in *Information Wars as a Struggle Between Geopolitical Opponents Civilizations and Ethos*. Collection of Words of All-Russian Scientific Conference, Novosibirsk, April 26–27, 2018.

Khudoleev, B. V., "Informatsionnoye Protivoborstvo. Kogda Streliayut Slovom [Information Confrontation. When Shooting with Words]," *Krasnaya Zvezda [Red Star]*, No. 193, November 19, 2005.

Komov, S. A., "Informacionnaya Bor'ba v Sovremennoj Vojne: Voprosy Teorii [Information Struggle in Modern Warfare: Theoretical Issues]," *Voennaya Mysl' [Military Thought]*, 1996.

Kondiurin, V. I., and E. G. Tiutiunik, *Tekhnicheskie Sredstva Propagandy [Technical Means of Propaganda]*, Voennizdat, 1983.

Kondratyuk, M. O., "Informacijna Vijna ta Rol Mas-Media v Mizhnarodnyh Konfliktah [Information War and the Role of Mass Media in International Relations]," *Social'ni Komunikaciyi [Social Communications]*, Chapter II, Vistnyk [Journal], No. 41, 2013.

Korotchenko, E. G., "Informacionno-Psihologicheskoe Protivoborstvo v Sovremennyh Usloviah [Informational-Psychological Confrontation in Modern Conditions]," *Voennaya Mysl' [Military Thought]*, 1996.

Kostenko, N. I., "Mezhdunarodnaya Informatsionnaya Bezopasnost' v Ramkakh Mezhdunarodnogo Prava (Metodologiya, Teoriya) [International Information Security Within the Framework of International Law (Methodology, Theory)]," *Rossiyskiy Zhurnal Pravovykh Issledovaniy [Russian Journal of Legal Research]*, Vol. 5, No. 4 (17), 2018.

Kovtun, Yu. V., "Zagrozi Informacijnomu Prostoru Derzhavi v Umovah Agresii Rosijs'koi Federacii [Threats to the State's Information Space Under the Conditions of the Aggression by the Russian Federation]," in *Current Problems Related to the Management of Information Security of the Government*, 8th Scientific and Applied Conference, Kyiv, Ukraine: Ukrainian Ministry of Education and Science, Institute for the Modernization of Educational Content of Ukraine, National Academy of the Security Service of Ukraine, and the Research Institute of Informatics and Law of the National Academy of Legal Sciences of Ukraine, 2017.

Kozhuk, Y., "Zametki Voyennogo Obozrevatelya: Arena Informatsionnogo Protivoborstva [Notes of a Military Observer: Information Confrontation Arena]," *Flag Rodiny [Flag of the Motherland]*, No. 124, 2002a.

———, "Zametki Voyennogo Obozrevatelya: Novoye Staroye Oruzhiye [Notes of a Military Observer: New Old Weapons]," *Flag Rodiny [Flag of the Motherland]*, No. 12, 2002b.

Krasnoslobodtsev, V., Y. Kuzmin, A. Raskin, and I. Tarasov, "Informatsionnaya Bor'ba kak Osnovnoy Atribut Sovremennoy Voyny [Information Warfare as the Primary Attribute of Modern War]," *Informatsionniye Voyny [Information Wars Journal]*, Vol. 3, No. 39, 2016.

Krysko, Vladimir Gavrilovich, *Sekrety Psikhologicheskoi Voiny [Secrets of Psychological War]*, Minsk, 1999.

Kugusheva, A., "Ot Informatsionnykh Voyn k Povedencheskim [From Information Wars to Behavioral Ones]," *Informatsionniye Voyny [Information Wars Journal]*, Vol. 1, No. 37, 2016.

Kuleshov, Y. E., V. V. Zhutdiev, and D. A. Fedorov, "Informatsionno-Psikhologicheskoye Protivoborstvo v Sovremennykh Usloviyakh: Teoriya i Praktika [Information-Psychological Confrontation in Modern Conditions: Theory and Practice]," *Vestnik Akademii Voennykh Nauk [Journal of the Academy of Military Sciences]*, No. 1, 2014.

Kuzio, Taras, and Paul D'Anieri, *The Sources of Russia's Great Power Politics: Ukraine and the Challenge to the European Order*, Bristol, U.K.: E-International Relations Publishing, June 25, 2018.

Lanoszka, Alexander, "Russian Hybrid Warfare and Extended Deterrence in Eastern Europe," *International Affairs*, Vol. 92, No. 1, 2016.

Larina, E., and V. Ovchinsky, "Novaya Voyennaya Strategiya SShA i Povedencheskiye Voyny [New U.S. Military Strategy and Behavioral Wars]," *Informatsionniye Voyny [Information Wars Journal]*, Vol. 3, No. 35, 2015.

Lata, V. F., V. A. Annenkov, and V. F. Moiseev, "Informatsionnoye Protivoborstvo: Sistema Terminov i Opredeleniy [Information Confrontation: System of Terms and Definitions]," *Vestnik Akademii Voennykh Nauk [Journal of the Academy of Military Sciences]*, No. 2, 2019.

Lenin, V. I., *The Impending Disaster and How to Deal with It [Grozyashaya Katastrofa I Kaka S nei Borotsa]*, 1917. As of May 15, 2021:
https://leninism.su/works/73-tom-34/1690-grozyashhaya-katastrofa-i-kak-s-nej
-borotsya.html

Levchenko, O. V., "Evolutsia Gibridnoi Vijiny Rosiys'koi Federatsii proty Ukraini [The Evolution of Russia's Hybrid War Against Ukraine]," *Nauka i Oborona [Science and Defense]*, No. 2, 2017.

Limno, A. N., and M. F. Krysanov, "Informatsionnoye Protivoborstvo i Maskirovka Voysk [Information Confrontation and Concealment of Forces]," *Voennaya Mysl' [Military Thought]*, No. 5, 2003.

Lipkan, V. A., "Suchasnyj Zmist Informacijnyh Operacij Proty Ukrayiny [An Accounting of Current Information Operations Against Ukraine]," *Aktual'ni Problemy Mizhnarodnyh Vidnosyn [Current Issues of International Relations]*, Pub. 102, Sec. 1, 2011.

Lysenko, Volodymyr, and Catherine Brooks, "Russian Information Troops, Disinformation, and Democracy," *First Monday*, Vol. 22, No. 5, May 7, 2018.

Lyutov, I. S., "Voennaya Nauka: Prioritetnye Napravleniia Razvitiia [Military Science: Priority Areas of Development]," *Informacionnyj Sbornik Shtaba Ob'edinennyh Vooruzhennyh Sil Gosudarstv-Uchastnikov Varshavskogo Dogovora [Information Digest of the Joint Armed Forces Headquarters of the State Parties to the Warsaw Pact]*, June 30, 1990.

Makarenko, Ye. A., "Informacijne Protyborstvo u Suchasnyh Mizhnarodnyh Vidnosynah [Information Confrontation in Modern International Relations]," *Mizhnarodni Vidnosyny, Seriya «Politychni Nauky» [International Relations, Political Science Series]*, No. 17, 2017.

Makarov, D., "Informationnyye Voyny. Slovo, Postavlennoye pod Ruzh'ye [Information Wars. The Word, Placed at Gunpoint]," *Flag Rodiny [Flag of the Motherland]*, No. 115, July 4, 2009.

Markov, A., "Informatsionnoye Obespecheniye [Information Support]," *Armeiskii Sbornik [Army Digest]*, No. 11, November 1997.

Materialy dlia Dokladov po Voennoi Propagande [Military Propaganda Reporting Materials], Krymgosizdat, 1928.

Matveeva, E. Y., and I. V. Nosova, "Informatsionniy Extremism: Sushnost' i Proyavleniya [Information Extremism: Characteristics and Manifestations]," in *Informatsionniye Voyny kak Bor'ba Geopoliticheskikh Protivnikov, Tsivilizatsii i Razlichnykh Etosov [Information Wars as Struggle Between Geopolitical Opponents Civilizations and Ethos: Collection of Works of All-Russian Scientific Conference]*, Novosibirsk: Siberian State University Telecommunications and Information, April 26–27, 2018, pp. 434–443.

Matvienko, Yu., "'Tsvetniye' Revoliutsii kak Nevoyenniy Sposob Dostizheniya Politicheskikh Tseley v Gibridnoy Voyne: Sushnost', Soderzhaniye, Vozmozhniye Mery Zashity i Protivodeystviya ['Color' Revolutions as Non-Military Means to Achieve Political Goals in 'Hybrid' War: Nature, Content, Possible Protection Measures and Countermeasures]," *Informatsionniye Voyny [Information Wars Journal]*, Vol. 4, No. 40, 2016.

Matvienko, Yu. A., "Nevoyennyye Ugrozy kak Sostavnaya Chast' Sovremennogo Mezhgosudarstvennogo Protivoborstva [Non-Military Threats as a Part of Contemporary Interstate Confrontation]," *Vestnik Akademii Voennykh Nauk [Journal of the Academy of Military Science]*, Vol. 1, No. 58, 2017.

McDermott, Roger, "Gerasimov Unveils Russia's 'Strategy of Limited Actions,'" *Real Clear Defense*, March 11, 2019.

Medvedev, Dmitriy, and Asiyat Tarchokova, "Instrumenty Mezhdunarodnoy Legitimatsii Vneshnepoliticheskikh Deystviy Rossii v Usloviyakh Informatsionnogo Protivoborstva [Instruments for the International Legitimation of Foreign Political Actions in Russia in the Conditions of Information Confrontation]," *Informatsionnyye Voyny [Information Wars]*, Vol. 47, No. 3, 2018.

Memorandum of Policy No. 30, "Command and Control Warfare," Washington, D.C.: Chairman of the Joint Chiefs of Staff, Joint Staff, July 17, 1990, revised March 8, 1993.

Menchikov, Peter, "Information Battlefield [Informacionnoye Pole Boya]," *Nacionalnaya Oborona*, November 2020.

Ministry of Defense of the Russian Federation, "Asimmetrichnyye Voyennyye (Boyevyye) Deystviya [Asymmetric Military (Combat) Activities]," in *Spravochnik Po Terminologii v Oboronnoy Sfere [Defense Terminology Repository]*, undated a. As of May 20, 2021:
http://dictionary.mil.ru/folder/123102/item/129185/

———, "Informatsionnaya Voyna [Information War]," *Voyennyy Entsiklopedicheskiy Slovar' [Military Encyclopedic Dictionary]*, translated by Polina Kats-Kariyanakatte, Joe Cheravitch, Clint Reach, undated b. As of November 10, 2020:
https://encyclopedia.mil.ru/encyclopedia/dictionary/details.htm?id=5211
@morfDictionary

————, "Informatsionnoe Protivoborstvo [Information Confrontation]," *Voyennyy Entsiklopedicheskiy Slovar' [Military Encyclopedic Dictionary]*, undated c. As of November 10, 2020:
https://encyclopedia.mil.ru/encyclopedia/dictionary/details.htm?id=5221 @morfDictionary

————, *Kontseptual'nye Vzglyady na Deyatel'nost' Vooruzhennykh Sil Rossiyskoy Federatsii v Informatsionnom Prostranstve [Conceptual Views on the Activities of the Armed Forces of the Russian Federation in the Information Space]*, 2011, section 1. As of July 8, 2020:
http://ens.mil.ru/science/publications/more.htm?id=10845074@cmsArticle

————, "Maskirovka [Concealment]," *Voyennyy Entsiklopedicheskiy Slovar' [Military Encyclopedic Dictionary]*, undated d. As of January 21, 2021:
https://encyclopedia.mil.ru/encyclopedia/dictionary/details.htm?id=13494 @morfDictionary

————, "Ministr Oborony Sergey Shoygu Nazval Glavnoy Tsel'yu Informatsionnoy Voyny Zapada Protiv Rossii Polnoye Yemu Podchineniye [Defense Minister Sergey Shoygu Called Complete Submission to the West as the Main Goal of the Information War of the West Against Russia]," June 26, 2019. As of November 13, 2020:
https://function.mil.ru/news_page/country/more.htm?id=12238562@egNews

————, "Psikhologicheskaya Bor'ba [Psychological Struggle]," *Voyennyy Entsiklopedicheskiy Slovar' [Military Encyclopedic Dictionary]*, undated e. As of January 21, 2021:
https://encyclopedia.mil.ru/encyclopedia/dictionary/details.htm?id=9563@ morfDictionary

Ministry of Foreign Affairs of the Russian Federation, *Foreign Policy Concept of the Russian Federation*, November 30, 2016.

Ministry of the Interior of the Russian Federation, "Sredstva i Sposobi Informatsionnogo Vozdeystviya v Sovremennom Mire [Means and Methods of Information Impact in the Modern World]," Saint Petersburg University, Department of Special Information Technologies, April 30, 2020. As of January 21, 2021:
213.182.177.142/kafedr/19.Special'nih_informacionnih_tehnologii/OIB_FZO_19/ JI_3_1/JI_3_1.htm

Modestov, S. A., "Strategicheskoye Sderzhivaniye na Teatre Informatsionnogo Protivoborstva [Strategic Containment in the Theater of Information Confrontation]," *Vestnik Akademii Voennykh Nauk [Journal of the Academy of Military Sciences]*, No. 1, 2009.

Modestov, S. A., D. A. Nikitin, and E. A. Rabchevsky, "Sotsial'niye Seti kak Teatr Informatsionnogo Protivoborstva v Usloviyakh Sovremennoy 'Gibridnoy' Voyni [Social Networks as a Theater of Information Confrontation in Today's Hybrid War]," *Vestnik Akademii Voennykh Nauk [Journal of the Academy of Military Sciences]*, No. 3, 2019.

Molchanov, N. A., "Informatsionniy Potentsial Zarubezhnykh Stran kak Istochnik Ugroz Voyennoy Bezopasnosti RF [Information Potential of Foreign Countries as a Source of Threats to Military Security of the Russian Federation]," *Voennaya Mysl' [Military Thought]*, No. 10, 2008.

Monaghan, Andrew, "How Moscow Understands War and Military Strategy," *CNA*, November 2020.

Morozov, Yu., "Primeneniye SShA 'Miagkosilovogo Arsenala' v Sovremennom Mire [Application of 'Soft Power Arsenal' by the USA in the Modern World]," *Informatsionniye Voyny [Information Wars Journal]*, Vol. 1, No. 41, 2017.

Muhin, V., and I. Rekunko, "Ispol'zovaniye Smartfonov—Ugroza Utechki Konfidentsial'noy Informatsii [The Use of Smartphones—the Threat of Leaking Confidential Information]," *Informatsionniye Voyny [Information Wars Journal]*, Vol. 4, No. 44, 2017.

Mushinsky, Mikhail, "Strategii, Kontseptsii, Doktriny v Pravovoy Sisteme Rossiyskoy Federatsii: Problemy Statusa, Yuridicheskoy Tekhniki i Sootnosheniya Drug s Drugom [Strategies, Concepts, Doctrines in the Legal System of the Russian Federation: Problems of Status, Legal Technique and Interrelation]," *Yuridicheskaya Tekhnika [Legal Technique]*, No. 9, 2015.

"Na Informatsionnom Fronte, Est' li u Miloshevicha Svoi Udugov? [On the Information Front, Does Milosevic Have Own Udugov?]," *Soldat Otechestva [Soldier of the Motherland]*, No. 39, 1999.

National Security Council of the Russian Federation, *Osnovy Gosudarstvennoy Politiki Rossiyskoy Federatsii v Oblasti Mezhdunarodnoy Informatsionnoy Bezopasnosti na Period do 2020 Goda [Basic Principles for the Russian Federation's State Policy in the Field of International Information Security to 2020]*, July 24, 2013. As of October 29, 2020:
http://www.scrf.gov.ru/security/information/document114/

"Natsional'naia Bezopasnost', Informatsiia—Tozhe Oruzhie [National Security: Information Is Also a Weapon]," *Vestnik Voennoi Informatsii [Journal of Military Information]*, No. 10, 1998.

Naumov, Alexey, "Russkii Otvet: Chevo Rossiiya Dobilas Za Chetire Goda Voini V Sirii [Russian Answer: What Russia Has Achieved in Four Years of War in Syria]," *Lenta.ru*, September 30, 2019. As of May 15, 2021:
https://lenta.ru/articles/2019/09/30/syria/

Nemtsova, Anna, "Why Is Ukraine's War So Bloody? The Soviet Union Trained Both Sides," *The Daily Beast*, April 14, 2017.

———, "Putin's Crackdown on Dissent Is Working," *The Atlantic*, March 22, 2019.

Nogovitzin, A., "Informatsionnaia Voyna: Novii Vizov Budushchego [Information War: A New Future Challenge]," *Armeiskii Sbornik [Army Digest]*, No. 4, April 2009.

Novik, A., "Stavka na Spetsoperatsii [Bet on Special Operations]," *Strazh Baltiki [Guardian of the Baltic]*, No. 104, 2009.

Nuzhdin, O., "Informatsionniye Voyni XV Veka: Preliudiya k Sovremennosi [Information Wars of the XV Century: Prelude to Modernity," in *Informatsionniye Voyny kak Bor'ba Geopoliticheskikh Protivnikov, Tsivilizatsii i Razlichnykh Etosov [Information Wars as Struggle Between Geopolitical Opponents Civilizations and Ethos: Collection of Works of All-Russian Scientific Conference]*, Novosibirsk: Siberian State University Telecommunications and Information, April 26–27, 2018, pp. 517–527.

Nuzhdin, Y., "Informatsionniye Voyny: Uroki Devianostykh [Information Wars: Lessons of the Nineties]," *Flag Rodiny [Flag of the Motherland]*, November 22, 2000.

Oliker, Olga, Lynn E. Davis, Keith Crane, Andrew Radin, Celeste Gventer, Susanne Sondergaard, James T. Quinlivan, Stephan B. Seabrook, Jacopo Bellasio, Bryan Frederick, Andriy Bega, and Jakub P. Hlavka, *Security Sector Reform in Ukraine*, Santa Monica, Calif.: RAND Corporation, RR-1475-1-UIA, 2016. As of May 10, 2021:
https://www.rand.org/pubs/research_reports/RR1475-1.html

Orlansky, V. I., "Informatsionnoye Oruzhiye i Informatsionnaya Bor'ba: Real'nost' i Domysly [Information Weapons and Information Warfare: Reality and Speculation]," *Voennaya Mysl' [Military Thought]*, No. 1, 2008.

"Osnovniye Napravleniya Obespecheniya Informatsionnoy Bezopasnosti v Deyatel'nosti Voysk (Sil) [Main Trends in the Information Security of Troops (Forces)]," *Boevaya Vakhta [Battle Watch]*, No. 99, December 22, 2001.

Ovchinnikov, V. V., and N. M. Petrovich, "Information Confrontation in Modern Geopolitics [Informacionnoye Protivoborstvo V Covrememmoi Geopolitike]," Russian Ministry of Internal Affairs Joint Editing Office, undated. As of May 15, 2021:
http://pda.ormvd.ru/pubs/238/15196/

Palitay, I., "Vliyaniye Sredstv Massovoy Informatsii na Protsess Politicheskogo Vospriyatiya [The Influence of the Media on the Process of Political Perception]," *Informatsionniye Voyny [Information Wars Journal]*, Vol. 1, No. 45, 2018.

Panarin, I., "Sistema Informacionnogo Protivoborstva [System of Information Confrontation]," *Voenno-Promyshlennii Kur'er [Military-Industrial Courier]*, No. 14, October 14, 2008.

Partiino-Politicheskaia Rabota v Boevoi Obstanovke [Political and Party Work in a Combat Situation], Voennizdat, 1940.

Petrov, Ivan, "Shoigu Announced the Creation of Information Operations Troops [Shoigu Obyavil O Sozdanii Voisk Informacionnih Operacii]," *Rossiskaya Gazeta*, February 22, 2017. As of May 20, 2021:
https://rg.ru/2017/02/22/shojgu-obiavil-o-sozdanii-vojsk-informacionnyh-operacij.html

Pievtsov, H., A. M. Hordiienko, S. V. Zalkin, S. O. Sidchenko, A. O. Feklistov, and K. I. Khudarkovskyi, *Informatsiino-Psykholohichna Borotba u Voiennii Sferi [Information and Psychological Warfare in the Military Domain]*," Kharkiv, Ukraine: Kharkiv National University, 2017.

Pirumov, V. S., and M. A. Rodionov, "Nekotorye Aspekty Informacionnoj Bor'by v Voennyh Konfliktah [Some Aspects of Information Struggle in Military Conflicts]," *Voennaya Mysl' [Military Thought]*, No. 5, 1997.

Podvigin, E., "Gosudarstvenno-Vlastnaya Elita v Usloviyah Informatsionnogo Obshchestva: Vyzovy i Perspektivy [State Power Elite in the Information Society: Challenges and Prospects]," *Informatsionniye Voyny [Information Wars Journal]*, Vol. 1, No. 45, 2018.

"Politychni Komunikaciyi za Umov Mizhderzhavnyh Konfliktiv [Political Communications for the Purpose of Interstate Conflict]," *Analytical Report Based on the Results of an International Roundtable*, Kyiv, Ukraine: Taras Shevchenko University of Kyiv, Institute of International Relations, March 30, 2015.

Polyakova, Alina, "The Kremlin's Latest Crackdown on Independent Media," *Brookings*, December 6, 2017.

Pozdnyakov, A. I., "Informacionnaya Bezopasnost' Lichnosti, Obshestva, Gosudarstva [Information Security of People, Society, and the State]," *Voennaya Mysl' [Military Thought]*, No. 10, 1993.

"Pozdravliaem Yubilyarov [Happy Anniversary]," *Voennaya Mysl' [Military Thought]*, No. 12, December 2017.

Prudnikov, D. P., "Gosudarstvennaya Informatsionnaya Politika v Oblasti Oborony: Iskhodnoye Opredeleniye [State Information Policies in the Defense Sphere: Initial Definition]," *Voennaya Mysl' [Military Thought]*, Vol. 3, 2008.

Prysiazhniuk, D., "Zastosuvannya Manipuliatyvnykh Psykhotekhnolohii z Boku Rosii v ZMI Ukrainy (Na Prykladi Krymu) [Application of Manipulative Psychotechnologies by Russia in Ukrainian Media (on the Example of Crimea)]," *Visnyk Kyyivs'koho Natsional'noho Universytetu Imeni Tarasa Shevchenka: Viys'kovo-Spetsial'ni Nauky [Journal of the Taras Shevchenko National University of Kyiv: Military-Special Sciences]*, No. 23, 2009.

"Putin Nazval Klycheboi Temy Ukrepleniya Rossiskoi Grazdanskoi Identichnosti [Putin Called Strengthening Russian Civic Identity to Be Key Theme]," *RBC*, March 30, 2021. As of May 15, 2021: https://www.rbc.ru/rbcfreenews/6063198f9a794794c0383e43

Puzenkin, I. V., and V. V. Mikhailov, "Rol' Informacionno-Psihologicheskih Sredstv v Obespechenii Oboronosposobnosti Gosudarstva [The Role of Informational-Psychological Means in Ensuring the Defense of the State]," *Voennaya Mysl' [Military Thought]*, Vol. 24, No. 3, 2015.

Pyastolov, Sergey M., "Slovo Kak Oruzhiye [Word as a Weapon]," *Informatsionniye Voyny [Information Wars Journal]*, Vol. 1, No. 49, 2019.

Radio Free Europe/Radio Liberty, "Investigative Report: On the Trail of the 12 Indicted Russian Intelligence Officers," July 19, 2018. As of January 25, 2021: https://www.rferl.org/a/investigative-report-on-the-trail-of-the-12-indicted-russian -intelligence-officers/29376821.html

Ramus', Vladimir, "Osobiy Front [Special Front]," *Suvorovskii Natisk [Suvorov Onslaught]*, No. 16, 2019.

Raskin, A., and I. Tarasov, "Informatsionnoye Protivoborstvo v Sovremennoy Voyne [Information Confrontation in Modern Warfare]," *Informatsionniye Voyny [Information Wars Journal]*, Vol. 4, No. 32, 2014.

Rastorguyev, S. P., "An Introduction to the Formal Theory of Information Warfare," Moscow, 2003, cited in Timothy L. Thomas, *Comparing US, Russian, and Chinese Information Operations Concepts*, Fort Leavenworth, Kan.: Foreign Military Studies Office, February 2004, pp. 6–9.

"Reckless Campaign of Cyber Attacks by Russian Military Intelligence Service Exposed," *National Cyber Security Centre*, October 3, 2018.

Repko, S. I., "Voyna i Propaganda [War and Propaganda]," *Novosti [The News]*, 1999.

Rodionov, M. A., "K Voprosu o Formakh Vedeniya Informatsionnoy Bor'by [On the Question of the Ways of Waging Information Warfare]," *Voennaya Mysl' [Military Thought]*, Vol. 2, No. 2, 1998.

Romanov, M. S., "The Participation of the Russian Federation's Scientific and Educational Institutions in Special Information Operations." In *Current Problems Related to the Management of Information Security of the Government*, 8th Scientific and Applied Conference, Kyiv, Ukraine: Ukrainian Ministry of Education and Science, Institute for the Modernization of Educational Content of Ukraine, National Academy of the Security Service of Ukraine, and the Research Institute of Informatics and Law of the National Academy of Legal Sciences of Ukraine, 2017.

Romashkina, Nataliia, "Informatsionnyy Suverenitet v Sovremennuyu Epokhu Strategicheskogo Protivoborstva [Information Sovereignty in the Contemporary Age of Strategic Confrontation]," *Informatsionnyye Voyny [Information Wars]*, Vol. 52. No. 4, 2019.

Rossiyskiy Ekonomicheskiy Universitet Imeni G. V. Plekhanova [G. V. Plakhanov's Russian Economic University], "Voyennaya Sluzhba v 6 Nauckhoy Rote 8 Upravleniya General'nogo Shtaba Vooruzhennykh Sil Rossiyskoy Federatsii v Krasnodarskom Vysshem Voyennom Uchilishe Imeni Generala Armii S. M. Shtemenko [Military Service in the 6th Research Company of the 8th Directorate of the General Staff of the Armed Forces of the Russian Federation in the Army General S. M. Shtemenko's Krasnodar Higher Military School]," Enclosure 3 "List of Specialties," undated. As of January 25, 2021: https://www.rea.ru/ru/events/Pages/nauch-armia.aspx

Russian Federation, *Doktrina Informatsionnoy Bezopasnosti Rossiyskoy Federatsii [Information Security Doctrine of the Russian Federation]*, December 5, 2016.

————, *Federal Law No. 172-FZ, About Strategic Planning in the Russian Federation*, June 28, 2014a.

————, "Russian National Security Blueprint," *Rossiiskaya Gazeta [Russian Gazette]*, December 26, 1997.

————, *Russian National Security Concept*, January 10, 2000.

————, *Strategiya Natsional'noy Bezopasnosti Rossiyskoy Federatsii do 2020 Goda [National Security Strategy of the Russian Federation Until 2020]*, May 2009.

————, *Strategiya Natsional'oi Bezopsnosti Rossiyskoy Federatsii do 2020 Goda [National Security Strategy of the Russian Federation Until 2020]*, December 2015.

————, *Voyennaya Doktrina Rossiyskoy Federatsii [The Military Doctrine of the Russian Federation]*, December 25, 2014b.

Ryabchuk, V., and V. Nichipor, "Prognozirovaniye i Predvideniye v Sisteme Planirovaniya Operatsii i Obshchevoyskovogo Boya [Forecasting and Prediction in Operational Planning Systems and Combined Arms Combat]," *Armeiskii Sbornik [Army Digest]*, No. 10, October 2012.

Samokhvalov, V. I., "Spetsifika Sovremennoy Informatsionnoy Voyny: Sredstva i Tseli Porazheniya [Specifics of Modern Information War: Means and Purpose of Damage]," *Filosofiya i Obshestvo [Philosophy and Society]*, No. 3, July–September 2011.

Sayfetdinov, K. I., "Informatsionnoye Protivoborstvo v Voyennoy Sfere [Information Confrontation in the Military Sphere]," *Voennaya Mysl' [Military Thought]*, No. 7, 2014.

Selivanov V. V., and Y. D. Ilin, "Metodicheskiye Osnovi Formirovaniya Asimetricheskih Otvetov v Voyenno-Tekhnicheskom Protivoborstve s Visokoteknologichnim Protivnikom [Methodological Foundations for Forming Asymmetric Responses in Military-Technical Confrontation with a High-Technological Adversary]," *Voennaya Mysl' [Military Thought]*, February 1, 2019.

————, "Metodika Kompleksnoy Podgotovki Asimmetrichnykh Otvetov pri Programmno-Tselovom Planirovanii Razvitiya Vooruzheniya [Methodological Foundations for Forming Asymmetric Responses in Enterprise Planning for Arms Development]," *Voennaya Mysl' [Military Thought]*, February 1, 2020.

Semchenkov, A. S., "Protivodeystviye Sovremennym Ugrozam Politicheskoy Stabil'nosti v Sisteme Obespecheniya Natsional'noy Bezopasnosti Rossii [Responding to Modern Threats to Political Stability in Russia's National Security System]," dissertation, Moscow Federal University of M. V. Lomonosov, 2012.

Shehovtsov, N., and Y. Kuliashou, "Informatsionnoye Oruzhiye: Teoriya i Praktika v Informatsionnom Protivoborstve [Information Weapon: Theory and Application in Information Confrontation]," *Vestnik Akademii Voennykh Nauk [Journal of the Academy of Military Sciences]*, Vol. 1, No. 38, 2012.

Shevcov, A. A., "Information Strategy on the Russian Federation Based on the Example of the Military Conflict in Syria," *Communicology: Electronic Scientific Magazine*, Vol. 3, No. 1, 2018, pp. 59–67.

Shevtsov, V. S., "Informatsionnoye Protivoborstvo v Globaliziruyushemsia Mire: Aktual'nost', Differentsiatsiya Poniatiy, Ugrozy Politicheskoy Stabil'nosti [Information Confrontation in a Globalizing World: Relevance, Differentiation of Concepts, Threats to Political Stability]," *University Journal [Vestnik Universiteta]*, No. 5, 2015.

Shil'bakh, K., and V. Sventsitsii, *Voennye Razvedki [Military Intelligence]*, Voennoe tipografnoe upravlenie, 1927.

Shitov, Andrei, "Dve Doktrini: Chem Otlichajutsa Podhodi Rasii u SSHA k Informacionnoi Bezopastnosti [Two Doctrines: What Are the Differences Between the Russian and U.S. Approaches to Information Security]," *TASS*, December 12, 2019.

Sinchuk, Yu., "Sposoby Vedenija Sovremennyh Vojn [Methods of Conducting Modern Wars]," *Voennaya Mysl' [Military Thought]*, 2000.

Sirotkin, D., A. Tyrtyshny, and A. Makarenkov, "Model' Pravovogo Regulirovaniya v Oblasti Informatsionnogo Protivoborstva [Model of Legal Regulation in the Field of Information Confrontation]," *Informatsionnyye Voyny [Information Wars]*, No. 3, 2016.

Slipchenko, V., "Novaya Forma Bor'by. V Nastupivsheme Veke Rol' Informatsii v Beskontaktnykh Voynakh Budet Lish' Vozrastat' [A New Form of Combat. In the Coming Century the Role of Information in the Contactless Wars Will Only Increase]," *Armeiskii Sbornik [Army Digest]*, No. 12, 2002.

———, "Informatsionnyy Resurs i Informatsionnoye Protivoborstvo [Information Resources and Information Confrontation]," *Armeiskii Sbornik [Army Digest]*, No. 10, 2013.

Smith, Charles F., "Command, Control and Communications Countermeasures (C3CM)," *Military Review*, 1983.

Soldatov, Andrei, and Irina Borogan, "Russia's Approach to Cyber: The Best Defence Is a Good Offence." In *Hacks, Leaks and Disruptions: Russian Cyber Strategies*, European Union Institute for Security Studies, October 1, 2018, pp. 15–24.

Soldatov, Andrei, and Michael Rochlitz, "The *Siloviki* in Russian Politics," in *The New Autocracy: Information, Politics, and Policy in Putin's Russia*, ed. Daniel Treisman, Washington, D.C.: The Brookings Institution, 2018.

Stavickii, A. V., "Ontologicheskiye Osnovy Informatsionnoy Voyny v Kontekste Bolshoy Igry Protiv Rossii [Ontological Foundations of Information Warfare in the Context of the Great Game Against Russia]," in *Informatsionniye Voyny kak Bor'ba Geopoliticheskikh Protivnikov, Tsivilizatsii i Razlichnykh Etosov [Information Wars as Struggle Between Geopolitical Opponents Civilizations and Ethos: Collection of Works of All-Russian Scientific Conference]*, Novosibirsk: Siberian State University Telecommunications and Information, April 26–27, 2018, pp. 664–677.

Streltzov, A. A., "Osnovniye Zadachi Gosudarstvennoy Politiki v Oblasti Informatsionnogo Protivoborstva [Primary Issues for Government Policies in the Area of Information Confrontation]," *Voennaya Mysl' [Military Thought]*, 2013.

Suleimanova, Sh. S., and E. A. Nazarova, *Informatsionniye Voyny: Istoriya i Sovremennost' [Information Wars: Past and Present]*, Moscow, 2017.

Suvortseva, Elena, "Voiny Budushchego. Zapad Perekhodit v 'Informatsionnoe Nastuplenie' [Wars of the Future. The West Starts an 'Information Attack']," *Na Boevom Postu [At the Fighting Post]*, No. 42, 1997.

Taylor, Margaret L., "Combatting Disinformation and Foreign Interference in Democracies: Lessons from Europe," *Brookings*, July 31, 2019.

Thomas, Timothy L., "Russian Views on Information-based Warfare," *Airpower Journal*, July 1996.

———, *Russia Military Strategy: Impacting 21st Century Reform and Geopolitics*, Fort Leavenworth, Kan.: Foreign Military Studies Office (FMSO), 2015.

Torsukov, Evgenii, "Pravnuk Bismarka Sotrudnichal s Krasnoy Armiey [Bismarck's Great-Grandson Collaborated with the Red Army]," *Nezavisimmoe Voennoe Obozrenie [Independent Military Review]*, No. 27, 2003.

Tritten, James J., *Military Doctrine and Strategy in the Former Soviet Union: Implications for the Navy*, Monterey, Calif.: Naval Postgraduate School, 1993.

Trotsenko, K. A., "Informatsionnoye Protivoborstvo v Operativno-Takticheskom Zvene Upravleniya [Information Confrontation on the Operational-Tactical Level]," *Voennaya Mysl' [Military Thought]*, No. 8, 2016.

Tsygankov, A. M., "Voyenno-Politicheskiye Aspekty Stroitel'stva i Razvitiya Vooruzhennykh Sil Rossiyskoy Federatsii na Sovremennom Etape [Military and Political Aspects of Construction and the Development of the Armed Forces of the Russian Federation on the Modern Stage]," *Voyenno-Nauchnaya Konferentsiya Akademii Voyennykh Nauk [Military-Scientific Conference of the Academy of Military Sciences]*, Vol. 67, No. 2, 2019.

Turchenko, Fedir, and Halyna Turchenko, *Proiekt "Novorosiia" and Novitnia Rossiisko-Ukrainska Viyna [The Novorossiya Project and the Latest Russian-Ukrainian War]*, Kyiv, Ukraine: Institute of History of Ukraine, 2015.

Turovskii, Daniil, "Rossiyskie Vooruzhenye Kibersily kak Gosudarstvo Sozdaet Voennye Otryady Khakerov [Russian Armed Cyber Forces: How the State Creates Military Hacker Units]," *Meduza [Medusa]*, November 7, 2016.

"Ukraine's Defense Minister: 40% of Ministry's Officials Fail Polygraph Tests," *UNIAN Information Agency*, April 25, 2016.

U.S. Department of the Army, Headquarters, *FM 100-6 Information Operations*, Washington, D.C., August 1996.

Vayner, A. Ya., "O Protivoborstve v Sfere Upravlenia [On Confrontation in the Sphere of Command and Control]," *Voennaya Mysl' [Military Thought]*, No. 9, 1990.

Voennaia Propaganda v Armii: Materialy [Military Propaganda in the Army: Materials], Khar'kov izdanie ukrpura, 1921.

Volobuyev, E. I., "VMF i Problemy Kompleksnogo Ognevogo Porazhenia Protivnika [Navy and Problems of Defeating the Adversary through Complex Fires]," *Voennaya Mysl' [Military Thought]*, No. 4, 2003.

Vorob'yev, I. N., "Voprosy Teorii i Praktiki Manevrennoj Oborony [Questions of the Theory and Practice of a Mobile Defense]," *Voennaya Mysl' [Military Thought]*, No. 9, 1990.

Vorontsova, L. V., and D. B. Frolov, *Istoriya i Sovremennost' Informatsionnovo Protivoborstva [History and Modernity of Information Confrontation]*, Goryachaya Liniya-Telekom, 2006.

Yarkov, S. A., "Nevoyennyye Sredstva i Nevoyennyye Mery Neytralizatsii Voyennykh Opasnostey: Sushnostnoye Razlichiye i Predmetnaya Khrakteristika Poniatiy [Non-Military Means and Non-Military Measures to Neutralize Military Dangers: Essential Difference and Objective Characteristics of the Concepts]," *Natsional'naya Bezopasnost' [National Security]*, No. 3, 2017.

Yashchenko, Yu. O., "Internet i Informatrionnoye Protivoborstvo [Internet and Information Confrontation]," *Voennaya Mysl' [Military Thought]*, 2003.

Yavorska, H., "Hibrydna Viina yak Dyskursyvnyi Konstrukt [Hybrid Warfare as a Discursive Construct]," *Stratehichni Priorytety [Strategic Priorities]*, No. 4, 2016.

Yuzova, I., "Analiz Orhanizatsiyi ta Vedennya Informatsiyno-Psykholohichnykh Operatsiy Pry Vedenni Hibrydnoyi Viyny [Analysis of the Organization and Conduct of Informational-Psychological Operations in the Conduct of Hybrid Warfare]," *Zbirnyk Naukovykh Prats' Kharkivs'koho Natsional'noho Universytetu Povitryanykh Syl [Anthology of Research Works of Kharkiv National Air Force University]*, No. 2, 2020.

Zanovich, A. A., "Pol'skaia Razvedka Protiv Krasnoi Armii, 1920–1930-e gody [Polish Intelligence Against the Red Army, 1920–1930s]," *Voenno-Istoricheskii Zhurnal [Military and History Journal]*, No. 10, 2007.

Zavodovs'ka, O., "Formuvannya Informacijnoho 'Poryadku Dennoho' Yak Metod Vedennya Hibrydnoyi Vijny v Konteksti Suchasnyh Mizhnarodnyh Vidnosyn [Formation of the 'Daily Agenda' as a Method of Conducting Hybrid Warfare in the Context of Contemporary International Relations]," *Visnyk [Journal] of the Lviv University, International Relations Series*, Vol. 36, No. 3, 2015.

Zharkov, Vitalii, *Politicheskaia rabota v RKKA (1929–1939 gg.) [Political Work in the Red Army (1929–1939)]*, K. D. Ushinskii Yaroslavskii State Pedagogical University, 2005.

Zharkov, Y., ed., *Istoriia Informatsiino-Psykholohichnoho Protyborstva [History of Information and Psychological Confrontation]*, Kyiv, Ukraine: Research and Publishing Department of the National Academy of Security Service of Ukraine, 2012.

Zhdanova, Mariia, and Dariya Orlova, *Computational Propaganda in Ukraine: Caught Between External Threats and Internal Challenges*, Working Paper 2017.9, Computational Propaganda Research Project, Oxford University, 2017.

Zinchenko, Aleksandr, and Anastasiia Tolstukhina, "Mir ili Voyna v Kiberprostranstve? [Peace or War in Cyberspace?]," *Mezhdunarodnaya Zhizn' [International Life]*, No. 9, 2018.

Zinoviev, V., A. Koldunov, and N. Gruzdew, "Perspektivy Primeneniya Informatsionnykh Setey v Voyennom Dele [Possible Uses of Information Networks in Military Activities]," *Informatsionniye Voyny [Information Wars Journal]*, Vol. 1, No. 33, 2015.

Zolotuhin, V. M., "Sokhraneniye Sotsiokul'turnoy Rossiyskoy Identichnosti v Prostranstve Informatsionnykh Voyn [Preservation of Russian Sociocultural Identity in the Space of Information Wars]," in *Informatsionniye Voyny kak Bor'ba Geopoliticheskikh Protivnikov, Tsivilizatsii i Razlichnykh Etosov [Information Wars as Struggle Between Geopolitical Opponents Civilizations and Ethos: Collection of Works of All-Russian Scientific Conference]*, Novosibirsk: Siberian State University Telecommunications and Information, April 26–27, 2018, pp. 231–239.

Zorina, E., "Propaganda kak Sovremenniy Instrument Vozdeystviya na Obshestvennoye Soznaniye [Propaganda as a Modern Instrument of Influence on Public Opinion]," *Informatsionniye Voyny [Information Wars Journal]*, Vol. 4, No. 36, 2015.

Zushin, Evgenii Grigorevich, "Vlast', ne Imeyushchaia Ravnykh po Sile Vozdeystviia [Power, Unparalleled in Terms of Impact]," *Nezavisimoe Voennoe Obozrenie [Independent Military Review]*, No. 16, 1999.

Zvarych, A. O., "The Experience of Ukrainian Counteraction to the Negative Information and Psychological Influence of the Russian Federation," *Zbirnyk Naukovykh Prats' Kharkivs'koho Natsional'noho Universytetu Povitryanykh Syl [Collection of Scientific Works of Kharkiv National University of the Air Force]*, Vol. 56, May 22, 2018.

CPSIA information can be obtained
at www.ICGtesting.com
Printed in the USA
LVHW081403160922
728494LV00009B/219